The Classic Hewn-Log House

The Classic Hewn-Log House

A Step-by-Step Guide
to Building and Restoration

Charles McRaven

Storey Publishing

*The mission of Storey Publishing is to serve our customers
by publishing practical information that encourages personal
independence in harmony with the environment.*

Edited by Carleen Perkins

Cover designed by Wendy Palitz

Text design and layout by Jen Rork

Art Direction by Cynthia McFarland

Cover photographs by © Linda Moore McRaven: front cover top left, top right, and
bottom right, back cover top; Jonathan Stoke: back cover bottom left;
© Roger Wade/Oakbridge Timberframes: front cover bottom left

Interior photographs by © Linda Moore McRaven, except for the following:
© W. Cody/CORBIS 120; © William K. Geiger 125 bottom right, 128 top, 130 bottom
right, 134 top left; Jonathan Stoke 119, 121, 123 bottom, 129 top right, 133 top; Cindy
Thiede 123 top; © Roger Wade/Boyd Mountain Log Cabins 126 bottom, 128 bottom left
and right, 129 top left and 133 bottom; © Roger Wade /Carney Architects/On Site
Mangement/Rocky Mountain Log Homes 132 top; © Roger Wade/Pederson Logsmith
131; © Roger Wade/Yellowstone Traditions /Peace Designs/Candace Miller, A.I.A. 50, 127
top left, and 132 bottom left

Text production by Jennifer Jepson Smith

Line drawings by © Chandis Ingenthron and Charles McRaven

Indexed by Susan Olason

Library of Congress Cataloging-in-Publication Data

McRaven, Charles.
   The classic hewn-log house : a step-by-step guide to construction and restoration /
Charles McRaven.
     p. cm.
   Includes bibliographical references and index.
   ISBN 1-58017-590-2 (pbk. : alk. paper)
   1. Log cabins--Design and construction. 2. Historic buildings--Conservation and
restoration. I. Title.
TH4840.M34 2005
690'.837--dc22
2005004010

DEDICATION

To Linda, who has shared so many adventures with me, and to our five —
Amanda, Lauren, Chelsea, Ashley, and Charlie — growing up with the
legacy of the American log cabin, learning their building skills at my side.

ACKNOWLEDGMENTS

So many people helped with the new edition of this book. First, and always,
the guiding force behind my projects, my wife Linda — creating, supporting, organizing, photographing, editing, typing. Janet Pitt, our office com-
missar — always on top of it. Margaret Morris, our right hand — always
there to take the pressure off. Amanda, Lauren, Chelsea, Ashley, and Charlie, whose photos and energy clarified sticky text questions, and whose
great work at the National Building Museum (every year building a log
cabin) shared their enthusiasm with thousands of eager urban children.
The staff of Storey Publishing, whose work with me throughout the years
has always exceeded my high expectations — Deborah Balmuth, Carleen
Perkins, and Kent Lew. Art Thiede and the late Cindy Thiede for their
wonderful log books and photography. Cindy's talent, generosity, energy,
vision, and love of America's heritage will be missed. And, of course, Chandis Ingenthron, whose drawings after almost 30 years of working with us
keep getting better.

To all these wonderful folks, my deepest thanks.

CONTENTS

FOREWORD

FOR HUNDREDS OF YEARS, building with logs was the primary method of home construction in Europe and North America. The Dutch and Swedes introduced the log-building tradition to the Philadelphia area in the late 17th century, and it was the larger German population that disseminated the tradition, with the Scots-Irish helping to carry it to the frontier. By the time of the Revolution, it had become the dominant method of house construction in the colonies. In fact, a 1786 survey of 140 houses in the Shenandoah Valley of Virginia showed that one house was of frame construction, one was stone, and 138 were built from logs.

As early as 1803, a distinction was noted between a log cabin — a temporary dwelling built from rounded logs — and a log house, which was built from hewn or squared-off logs and was meant to be a permanent dwelling. Few, if any, original log cabins exist today, whereas a multitude of log houses, often embedded in later structures, dot the countryside.

Preserving the heritage of the hewn-log building tradition evokes a nostalgia for those dwellings of our forefathers; it is a vanishing art that needs to be preserved. Charles McRaven is a consummate craftsman with a vast knowledge of this art. His 50 years of experience are represented here in *The Classic Hewn-Log House*, the definitive guide for building, restoring, and preserving hewn-log structures. From choosing the site and design to selecting materials and demonstrating techniques, McRaven provides easy-to-follow directions written in a highly readable style that conveys the author's hands-on experience.

In this third edition, Charles McRaven has expanded on his lifelong study of the restoration of hewn-log dwellings, reinforcing his reputation as the country's leading authority on log construction and restoration. Just as Thomas Jefferson described Andrea Palladio's *I quattro libri dell'architettura* as the bible of classical architecture, one could say that McRaven's *The Classic Hewn-Log House* is the bible of hewn-log construction.

K. Edward Lay
Professor Emeritus of Architecture
University of Virginia

Hewn Logs and Houses

WITH THIS THIRD EDITION, it is now a quarter-century since the original writing, and we have built and restored and worked on hundreds of log houses all over the United States. The focus has shifted from hewing new wood in the mid-South to restoration and rebuilding in the East and other regions. But the principles are the same, the techniques true.

With so many available publications on building your own shelter, using everything from plywood to old car bodies, it seemed to me that a detailed work on traditional log houses was in order. Of course there are log cabin books in every bookstore, but these invariably deal with the round-log, "modern" cabins. Somehow, we have come to accept these pole cabins of sapwood logs, stacked schoolboy fashion, as the shelters America grew up in.

Not so. Historians and folklorists have established reliable records of the growth of log housing in this country, beginning not long after the settlement of Jamestown and Plymouth. With few exceptions, the pioneers built hewn-log houses.

Although its origins may be traceable to Scandinavian mead halls or old English half-timbered cottages, the squared-log homesteader's house of this country's uplands is an architectural form entirely its own.

The use of round logs was limited mostly to temporary cabins, barns, forts, and hastily thrown-up shelter. Our forebears built with much more skill to house their families. Hewing, adzing, notching, mortising, and riving — all these skills reached a high level in the building of a settler's permanent home.

And the labor involved was staggering. If there is a single reason hewn-log houses are so rarely built today, it is that just too much work is required. Everyone who attempts a log house today tries to avoid the very investment in painstaking, hard work that gave (and still gives) the hewn-log house its heart. Builders devise crack-brained schemes to shape logs mechanically into identical, fit-together games. The purveyors of log cabin kits peddle their wares to nostalgia buffs who want their houses assembled quickly, if not authentically.

The hewn-log house survived the seasons and the wear of generations because it was best. Best for the needs of the people, in best harmony with the land, the best use of the materials at hand. With its easily rotted sapwood hewn away, its corner notches locked rigid, its thick walls holding heat in or out, it stands as a beautiful part of our heritage, best because it has lasted so long and served so many .

Of course, today we can render those same massive logs into skinny 2×4s, plywood, fiberboard, particleboard, and fake paneling, and build several dwellings. I can only say that a log house I build will house many generations, as have those of the pioneers. The modern thin-walled, thin-roofed, thin-floored house may shelter one or two.

To build or restore a hewn-log house today, you must first come to terms with the materials and the way they must be worked. Study the settlers themselves; by understanding their way of life, you can come to understand their homes. Do not make the common mistake of assuming that you have the advantage of modern machines and technology. Or that, if done well with hand tools and limited knowledge in 1800, it will be done much better today.

The very use of machines has dulled our self-reliance, our inventiveness, our ability to think in harmony with the materials and with the earth. When you can learn to select the right wood for each purpose, knowing how it can be worked, how it responds to shaping, stress, seasoning, and dampness, you have only begun. You must then learn to cut that particular tree, hew or rive your timber, shape, smooth, fit it to its use, all by hand, before you begin to achieve the viewpoint necessary for this kind of work.

Then you must learn thousands of long-forgotten building pitfalls that were common knowledge to those same settlers. You must make mistakes serious enough to be remembered.

Not all of us will shape and drive a wooden peg equally well. Some of our forefathers, too, were terrible at building, and most of their flimsy efforts have fallen down. Those that remain are usually the best, and you can learn from them.

If you plan to build, start with something simple. Try shelves, a doghouse, or a small barn. You may in the process remove all doubt that you are indeed a bumbler, and only the cows will know.

Do not hurry. Almost all the problems modern owners of hewn-log houses encounter stem directly from their own frenzy to get inside. We are too conditioned to paying our money for instant gratification. Even the kit houses must be given time to settle, although their buyers may not be told this. Allow a year. Or two. Take time to savor this experience; you may not repeat it.

When I begin a hewn-log house, it does not matter that this is one of many I have raised or restored. The thrill of beginning a new affair returns, in the feel of worn tools in my hands, in the sight of new foundation stones rising from the ground. And the first long broadaxe strokes wake an old fever inside me. My hands harden again to the feel of wood and stone. And I look forward to the long months of building.

I tell first-time builders to begin in the fall, when the sap is down, when the crisp air makes handling the broadaxe a form of joy. Hew a few logs at a time — maybe one a week, or one a day. Work through the cold months. When the ground is too frozen to dig foundations, the big axe will have you in shirtsleeves. By spring you'll have a pile of hewn logs, partly seasoned,

lighter in weight, ready to be notched and raised on your new foundation stones.

Settling and shrinking will go on through the summer as you labor. Roof as soon as possible, but leave verticals like door and window facings, staircase uprights, and studs for the lean-to until a year has passed. Two years is better, but most of the settling will be over by the second fall. Chink last, when the logs are completely seasoned.

If you plan to restore, try to find a solid, aging pioneer cabin to move or restore on the spot. Or you may locate old hewn logs from a fallen house or a dismantled barn that can be bought and used to create your own house out of materials with their own history. Much of the work I do with logs today is with antique wood. The old axe marks always have a story to tell.

It has always been this tuning in to history that has moved me to work with logs. It is not hard to re-create the lives of the pioneers from the marks they left for us. The old wood tells softly of settlement, of taking new land, of the simple magic of creating a home. It is easy to understand the desire to extend the life of such an old house into your own.

The driving of the final wooden peg, the placing of the last chimney stone, is for me never an occasion of pure joy. It is inevitably an anticlimax — an aimless gathering of tools, leftover pieces in the unaccustomed quiet, a sad time of leave-taking. I become quite involved with my houses, to the extent that I am somewhat demanding when I restore or build for a client. It's something like the relationship an antique car enthusiast has with his treasures.

The chapters are arranged in a sequence to yield information about this very American house: its origins, types, and locations. Then the business of how to build or restore a log house is dealt with, from site selection to final details such as plumbing and utilities. In this revision, there is more emphasis on restoration. This is designed to be the source book as well as the handbook of hewn-log building and restoration.

I tell you about the evolution and growth of this type of house, its sizes, shapes, traditions. I tell, piece by piece, how to build one, and why it must be done this way. I have swung an axe since I was nine, and have made all the mistakes I warn of, rejecting systematically the methods that did not work.

I have built and restored many hewn-log houses over the years (and a few round-log ones) and there is no part of the work I do not do myself. I've forged nails and hinges, hewn logs and riven shakes, laid the stone and raised the beams by hand. I've gone into the forest with an axe and built from the standing trees.

I have dismantled, moved, rebuilt, and restored everything from corncribs to Victorian farmhouses hiding cores of hewn-log houses underneath. I do work to National Historic Register standards and I build workshops of logs. As long as we can be true to the material and the heritage, we continue the tradition.

My writing is from experience, from blisters earned and judgment sharpened in use. I do not work from someone else's plans. My tools are old tools, partly because I like old things and partly because, like the log houses themselves, old tools seem to work best.

But I learn many things anew each time I begin the labor of long love with the first foundation stone of another log house. I rediscover truths perhaps known by my ancestors, facing the forests of the frontier.

I must confess that, from the laying of the first stone for a log house, I work in close harmony with images of my own great-grandfathers. One, David Saxon, traveled from South Carolina in the 1850s to hew out a pine-log house among the low hills near Camden, Arkansas. Another, Daniel McGraw, left the cotton-sun of Mississippi in 1878 to settle a long slope of mountain above Arkansas's Mulberry River near Turner's Bend, in a cabin with a deerskin for a door.

And generations before them, west from Virginia and the Carolinas, the ghosts of earlier ancestors move through the forests on silent feet, axes on their shoulders, into the new land, up the lost creeks and the narrow hollows, sometimes to the very hillsides where I have hewn long after.

We may meet them, you and I, in some clear October dusk down in a beechwood glade, among the smell of clean wood and the bright chips scattered. They will look long at the work of our hands, measuring in the wisdom of their incredible years. And we will need no words.

Buck Mountain
Albemarle County, Virginia
January 2005

Roark Creek cabin, 1977

Roark Creek, 1992

Buck Mountain, 2005

CHAPTER ONE

The American House

THE SETTLER WHO SWUNG his long axe in the clearing to shape the logs for his pioneer home established a tradition curiously American: the log cabin. Even the term calls up mental pictures of open fireplaces, long rifles, and coonskin caps. It has launched presidential aspirations, and been the setting for a hundred years of hillbilly stories. But in this age of temporary housing, it retains its dignity.

Beginnings

The American log house is, by historic definition, a structure of hewn logs, corner-notched to form one or more pens, chinked with split boards or thin stones and mud or mortar, roofed on top with split shakes. It has one or more fireplaces, stone, brick, or mud-and-stick chimneys, and is intended as a permanent home.

The log cabin, by contrast, is often of round logs, and is of less-careful construction, being generally built as a temporary or occasional residence. Size has little to do with the basic difference between house and cabin, although this is the definition basis most people today would use — our "bigger is better" society speaking.

Cabin is from the Old French term *caban*. In *Pioneer America*, Eugene Wilson writes that the earlier "capanna," which may be the forerunner of these names, has Neolithic connections. American cabins have been mostly one-room structures, evolving from the "bay" or "rod" dimension — about 16 feet — of English rural housing.

The 16-foot measure, perhaps originally from the width required by four oxen, recurs again and again

This cabin restoration with stone and catted chimney is at Montebello, Virginia, built by the late black-powder gunsmith Rodney Harris.

throughout the growth and spread of log housing. That's largely cultural; the isolated pioneer as often measured his house using his three-foot axe handle. But the fact remains that logs of greater length were heavy and awkward for the lone settler to handle. His need was for quick shelter, so a small structure of the materials available — logs — filled that need best.

This settler often dreamed of a substantial country house like those of the gentry that rose along the rivers of the East. This log building would be temporary, and when his fields ran wide and the roads reached out from the teeming towns, he would build again. And the log house would be put aside for use by visiting relatives, or as servants' quarters — eventually even to be stuffed with hay.

D. A. Hutslar offers a historic comparison of log house versus log cabin in the journal *Ohio History*. Cabins, he notes, were of unhewn logs, chinked with rails and moss, straw and mud. Roofs were covered with long staves with weight poles, which are poles laid on the split roof shakes to hold them in place. There were no windows or chimneys. Log houses, on the other hand, were hewn, with stone and plaster in the chinks. The roofs were shingled; there were glass windows and chimneys.

It's a bit difficult to imagine a cabin with no chimney, yet the early ones often had no more than a hole in the roof for the smoke from the dirt-floor firepit to escape through. This was common in the peasant cottages of Britain. The gable-end fireplace so common in America developed in 15th-century England. Shortly thereafter, with the upper area now free of smoke, the use of loft space for living quarters became common. In central Europe, the chimney was usually in the center of the house.

The settlers from the tall ships anchored off Jamestown and Plymouth had been, for the most part, town dwellers. Although many were skilled carpenters with tools at hand, they had no knowledge of, or experience with, log construction. They set about reducing the formidable forest trees to whipsawn boards and riven clapboards to nail onto hewn timbers, just as they would have done in Europe. Housed temporarily in huts of sailcloth, branches, and thatch, they endured rain, cold, and Indian depredation while laboriously fashioning the kind of houses with which they were familiar.

They could have had snug, safe quarters of logs almost from the first, had they so chosen. Most historians are of the opinion that these immigrants just didn't know how to build with logs until the Swedes, Finns, and Germans brought their skills with them later in the 17th century.

Cabins today are being restored, often with added logs and modern additions. This one in Missouri was moved twice, with replacements both times.

However, because logs were used early for stockade walls, forts, and even jails, it's more likely that the first settlers clung to their complicated house-construction practices as a link with a culture they feared would soon fade into the wilderness. History is full of accounts of civilized people thrust into the wilds, clutching remnants of their ordered, familiar pasts.

The very persistence of hewn logs instead of round logs in the houses of the pioneers is as much a cultural matter as a practical one, given the relative labor and skill involved in building this way. Barns, corncribs, even temporary dwellings and hunters' shacks could be of round logs, but not the house in which the pioneer wife was to keep and raise her children. This house must have some pretensions to gentility, if only flat walls.

Remember, the wife was probably the moving force behind those pioneers-turned-planters, who rose from owning nothing but wild land to the ranks of the new gentry. And until she had her painted rooms and her plastered walls, she'd have them hewn as smooth as possible, thank you.

But of course the availability of so much timber, and soon the influx of log-wise Scandinavian and Middle European craftsmen, saw the log house emerge as the frontier structure. It could be built with only an axe if need be, and built well with an axe, auger, broadaxe, drawknife, hammer, and nails. No whipsawn boards, complicated mortising, or carefully finished woodwork went into the average frontier log house, although some finely crafted specimens were built.

Early Log House Construction

The settlers of New Sweden, on the Delaware River, are credited with the beginning of log house construction in America. Fort Christina, at what is now Wilmington, was built in 1638. In his book *The Log Cabin in America,* C. A. Weslager writes that half of the first settlers in New Sweden were Finns, whose building techniques were closer to the later styles here than those of the Swedes. Log houses were built inside a log stockade wall for protection, and soon others spread out into the countryside.

It takes some imagination today to envision this land blessed with straight trees that fit so well into the building traditions of these new Americans. Picture these settlers venturing farther and farther from the fort's walls to raise their log houses and establish their farms: up the Delaware, out into the waiting wilderness, the Indians, the wild game, the good land.

Some of these earlier dwellings were of round logs, some of hewn. Corner fireplaces were a Scandinavian feature different from either the German or Scots-Irish log houses soon to appear.

The influx of Germans into Pennsylvania and the advent of large numbers of Scots-Irish into the region by 1700 combined with the spread of the Scandinavian influence to create the log house as it has become known. Fred Kniffen, in his article "Folk Housing: Key to Diffusion," makes the significant statement that "Building with logs was a mode of construction, not an architectural type. Log, frame, stone or brick may all be the material for a type."

But it wasn't that simple: Corner-notched log pens just about mean square or rectangular houses or sections of houses, so the choice of types is somewhat limited. I do know some zealots who've built odd, multisided log houses in an attempt to be "different," but some other material would have been better. I think the materials — logs — influence type considerably. More about this later.

In *Folklore Today,* Warren Roberts discusses similarities in and differences between Scandinavian log houses and those in America, and concludes that there is little resemblance overall. Montell and Morse, in their book *Kentucky Folk Architecture,* write that the Pennsylvania Germans introduced the classic American log house. Henry Glassie, in "The Appalachian Log Cabin," states that "the log cabin stands as a symbol of this meshing of German and Scotch-Irish cultures."

We can safely say that the cultural building patterns of the Scandinavians, Germans, Scots-Irish, English, and even the Dutch underwent some necessary modifications to fit the conditions and materials available. The New Sweden houses were not of the careful ovallog, tightly fitted, chinkless style Roberts found where these people came from.

The Scots-Irish must have been overjoyed at tall, straight trees to build with, instead of the mud and stones and thatch of Ulster. And the Germans set to work building those wonderful barns of log and stone, along with their substantial American log houses with the Old Country touches in decorative beading, mitering, and even painting. In this land of trees, our ancestors naturally used them freely.

The spread of log housing followed the flow of settlers to new territory. Germans, traveling to the Shenandoah Valley in 1732, passed through Maryland, and of course many settled along the way. The Scots-Irish moved into western Virginia, too, and to the Carolinas. Ohio, Kentucky, Tennessee, and north Georgia were penetrated as the 1700s wore on. Even during the Revolution, thousands of land-hungry pioneers moved west, into the rich lands of the Tennessee River basin, fighting their own battles with Indians, the ever-encroaching forest, and the elements.

By 1800 the tide had reached into Alabama, where Eugene Wilson, in his book *Alabama Folk Houses,*

This dogtrot log house has entry doors off the breezeway. Its irregular roofline is a later addition. The dogtrot is a double-pen log house that has two separate rooms with usable space between the pens.

divides the folk house types into first, second, and third generations. He identifies first-generation houses with fine craftsmanship, built with skill, and made to endure. These were built until around 1840, when sawmills, commercial hardware, windows, and doors appear in the second generation. Less attention to craftsmanship is evident here. In the dogtrot house, Wilson notes a shift from the front door in each pen to entryways off the open passage during the period from 1840 to 1890. Third-generation houses, to 1940 or so, embody most of the modern rural building techniques.

Log House Styles

Mention of the dogtrot house brings us to the question of log house types or styles. Beginning with the basic one-room structure of one story, with gable-end chimney and door on the side parallel to the ridge, the types evolved in several directions.

A loft was the feature most often added, requiring perhaps three more courses of logs above the ceiling joists and a combination ceiling and upstairs floor. Given a relatively steep roof pitch, the living area was essentially doubled with the loft addition. These one-and-a-half-story, single-pen houses were built from the 1600s onward until perhaps the 1940s, when traditional log construction can be said to have ended.

That's a date out of the air, but it's close. Certainly during the Depression many hungry folks back from the city hacked out log houses for themselves on the folks' back forties. There, as before, that loft was the place the kids were stowed. Being a child of that period, I probably escaped sleeping in the loft of the cabin my father built only because the loft was never finished. We always had a lot of other projects to pursue.

Single- and Double-Pen

A full two-story, single-pen log house required only a few more logs, and these became common as permanent farmhouses, principally in the East and west into Ohio and Indiana. Like the loft house, the two-story required a peg ladder or narrow stair for reaching the upper floor.

Double-pen houses were logical expansions, given the weight of the logs. By building two separate pens, the settler could enlarge his house with easily managed short logs. He either built directly against his existing house at the gable end or set up the second pen some distance away and connected the two with one roof — the dogtrot.

A double-pen log house joined at the gable opposite the chimney is called just that, a double-pen house, although it can have other names. The double-pen house usually has two front doors, and a second chimney may be added at the gable end of the new pen. An example I am quite familiar with is the Beaver Jim Villines house at Ponca, Arkansas, preserved by the National Park Service as part of the Buffalo National River. The original pen logs are of hewn oak; the added pen is of cedar. Other houses are scattered throughout the South and East, some with both pens built at one time, some with the second pen added.

Saddlebag

When the added pen was joined at the chimney end, the house was called a saddlebag, putting the chimney in the center of the house. Sometimes the second pen was set up the chimney-depth apart from the original. In this case, there was usually a boxed-in passageway between the two pens. The chimney was often rebuilt or added to, to allow a fireplace to open into each pen. Double-pen houses of all kinds were often built all at once. The double fireplace is a good reason why, because it's harder to add a second chimney later in the center of the house. Sometimes the two pens shared the fourth interior log partition wall, which served to brace the full-length logs at midpoint.

Dogtrot

This practice of building all at one time was also, as often as not, the case with the dogtrot house, which has the two separate pens joined by a common roof. This type has become the classic log house, which allows a maximum of space for the logs used. It has become my favorite house.

Montell and Morse tend to believe both halves of the dogtrots they studied in Kentucky were built at one time. However, I know of as many dogtrot log houses with pens built at different times as I know of those built all together. Quite often a second pen would have different notching, indicating either another date or even transfer of the pen from another site. It should be pointed out that the more hard-nosed scholars insist that a true dogtrot have pens of

single-pen

saddlebag

saddlebag with shared wall

The one-and-a-half-story, single-pen log cabin was the basic pioneer structure in America. The saddlebag is a double-pen log house with a chimney between the pens. Often, the central chimney is open to both pens. The saddlebag does not have a usable room space between the two pens. When built at the same time, the saddlebag house often shared the fourth, internal log wall built with a complicated notch. It also usually has two front doors.

The Jacob Wolf dogtrot was used as an early courthouse, tavern, Indian agency, store, and dwelling in Norfolk, Arkansas. Its second-story logs are joined with triple notches.

identical dimensions and that the breezeway be floored. Because it is a colorful pioneer house type, modern rebuilders have certainly taken far-fetched liberties with non-dogtrot cabins to achieve the general effect. Many dogtrots have been closed in for more living space, and many have been boarded up entirely in complete camouflage

Wilson points out that the early Norse often utilized separate buildings for separate functions, and that these were often joined with a covered passage. Certainly the open-passage house was known prior to its appearance in this country as a log house.

The earliest example in the Ozarks region is the Jacob Wolf house in Norfolk, Arkansas, built in the 1820s. The dogtrot log house is generally considered a later development in this country, seen mostly in Alabama, Georgia, Kentucky, and Tennessee. The few examples of dogtrots in the older colonies on the East Coast are generally post–Civil War.

However, Terry Jordan, in *American Log Building*, notes that the Morton house, a now enclosed dogtrot log structure in Prospect Park, Pennsylvania, dates from 1690. Calling it an example of "Fenno-Scandia" building, he shows the unusual placement of the chimneys, on the inner-facing walls, off-center of the ridge. I have seen and read of examples in the Tennessee basin in the late 1700s. But whatever its origins, and whether built together or separately, the double-pen dogtrot house has become the stereotype of the substantial pioneer farmer's dwelling.

I consulted with the architect Tommy Jameson in 2000 on the restoration of the Jacob Wolf house. A mystery regarding dendrochronology dating came up — why were the logs cut over a six-year period? Close study showed the same craftsman hewed all the logs of this large house (left-handed hewing, which is my own natural approach although I'm right-handed). Whether it was Jacob Wolf himself, a slave, or Indian help, the same distinctive strokes were clear. We concluded that, with his blacksmithing skills, Indian agency work, homesteading, clearing, farming, hunting, it had just taken that long. Even such an ambitious dwelling apparently wasn't Jacob's highest priority.

Evolution of Styles

J. Frazer Smith, in his book *White Pillars*, sees the type as an evolutionary link between single-room log cabin and central-hall plantation house. With the passageway enclosed, and perhaps two rooms on each side of the hall thus created, you do have the basic planter's house of Tennessee. But the English Georgian houses of Sir Christopher Wren and their counterparts in Virginia had already set the pattern for the central–hall "big house" of the upland South.

A prime example of yet another form of the double-pen house, and somewhat of a riddle in itself, is the Whitaker-Waggoner log house, which stood in Morgan County, Indiana, before its removal to Indiana

University. This house was of two V-notched pens across the front, with a common log partition notched by means of a combination V- and half-notch, into the joined double-length logs of the front and back. In *American Folklife*, Warren Roberts dates the house reliably at about 1820, at which time apparently a rear wing, partially of frame and partially of log with single dovetail notching, was also built. Interlocking foundation stones place the coincidence of construction dates, and Roberts conjectures that perhaps the rear-log portion of the wing was moved from another site on the property. An old barn on the farm was of similar notching.

As a stonemason, I must point out here that interlocking foundation stones are not conclusive evidence of concurrent building. We often join foundation stones and duplicate mortar in our restoration work for an indistinguishable blend. In rebuilding and enlarging a slave cabin at Ash Lawn-Highland, President James Monroe's estate in Albemarle County, Virginia, we did this. We found the original quarry — from which the current owner let us take stone — and locked in our replacements with the originals. Few other stonemasons can detect the new work.

The Whitaker-Waggoner house showed a high degree of craftsmanship in several features. The joined interior log partition was rare, as was the continuous foundation, which could also have meant an earlier or original wing. Extended top gable-end logs supported the 18-inch-wide plates, which extended outward to form overhanging eaves. The rafters were fitted to these plates. The plates were kept from bowing out under the roof weight by timbers mortised and pinned between them, at each end of the front section and in the center over the partition. Roberts gives no details regarding a doorway between the upstairs rooms, because this timber would have spanned the building at the plate height of four feet. Perhaps a separate stair or ladder, not shown in the plans, gave access to the second upstairs room. This arrangement is common in eastern log houses.

This house also furnished convincing proof that many Indiana log houses, at least, were sided soon after construction. Clapboarding here went on before the end chimneys, and when removed showed little weathering of the logs underneath. This seems to have been a practice that rarely reached into the central states, although boards were quite often nailed over the logs in later years to keep out drafts and unwelcome fauna.

The late-1700s Page Meadows house in central Virginia had been sided at construction, with unmortared brick nogging between the logs. (Nogging is a clay filling used within hollow walls to increase mass and energy efficiency.) This is encountered in many Virginia houses, and is generally attributed to Thomas Jefferson's advocacy of the practice for "insulation" — actually more wall mass.

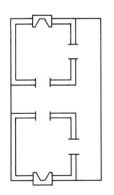

DOGTROT **The classic dogtrot floor plan of two separate pens, here showing a porch across the front.**

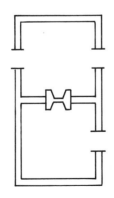

SADDLEBAG **Double chimney in the center was also typical of New England post-and-beam houses. Historians credit the central chimney to a German influence.**

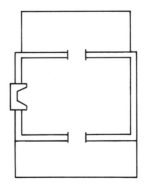

SINGLE-PEN LOG HOUSE **The basic one-room log cabin, shown with addition and porch floor plan. This was the departure point for most log cabin designs.**

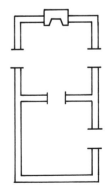

DOUBLE-PEN LOG HOUSE **This often had a second chimney at the opposite end. The two front doors were typical.**

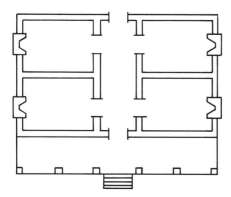

This is the plan of the mid-southern planter's house. Its central passage is also typical of the earlier Georgian houses.

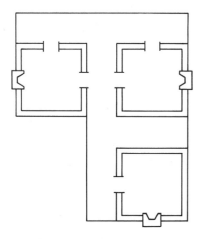

When additions were made to log houses, they were often separated by an open passage that was roofed over. This is a rear-wing addition to a dogtrot house, which has its own porch.

The mid-1700s Fischer brothers log house that stood near Winchester, Virginia, had its massive oak logs nicked to hold stucco, in another practice of covering the logs. Research by its restorer indicated that this house was built by the German Fischers for Lord Fairfax on land George Washington surveyed. The house was for an English family of new settlers named, coincidentally, Fisher. It was rebuilt near Slate Mills, Virginia, in the late 1980s.

The interior-partition log house, whether two full pens or with a common wall, is often referred to as a hall-and-parlor house. It has two rooms on the ground floor, and may or may not have a loft.

Two-story versions of the double-pen house fall into the "I" house type. This house is one room deep and two or more long, but is always two-story. It has gable-end chimneys, and in the early versions asymmetrical plans, according to Glassie in his book *Pattern in the Material Culture of the Eastern United States.*

Fred Kniffen calls the "I" house (not necessarily the log version) "the most widely distributed of all folk types" and sees it as the "symbol of economic attainment as a rural dwelling." The "I" plan often evolved as the earlier small farmers grew prosperous and wanted more pretentious dwellings. It is significant that this basic plan is also seen very early in French Louisiana, with a liberal use of galleries and French windows for circulation of air. Here the upper level became the main floor, with utility rooms brick- or tile-floored near ground level to contend with periodic flooding.

Additions

All the basic types of log houses were added to in other ways. The rear wing of the Whitaker-Waggoner log house, with its ridgeline at a right angle to that of the main house, is typical. This rear wing was usually aligned with one pen of the double-pen house to form an ell, and was almost always one-story. Similar wings were added to single-pen houses as well. Usually the rear unit was separate from the main house, because it was difficult to join logs to an existing structure. Another reason for a breezeway between the main house and rear wing was for fire protection, since the rear wing was often used as a kitchen.

A shed-roof lean-to was another addition common to all the basic log house types. It, too, often became a kitchen, but was invariably attached directly to the house. The lean-to was usually of rough-sawn lumber.

The lean-to wasn't limited to log houses. This logical appendage appears on frame houses, brick houses, and even barns. I do wish that whoever sold that brick-patterned tar paper, which is plastered over so many lean-tos, had stopped earlier.

Just as the log house itself emerged as the logical house type for the new country, peculiar local treatments evolved. Social custom and tradition mixed necessity, skill, and available tools and materials to produce local characteristics. As a general rule, with the passage of generations, less care was given to log

The log gable with roof purlins was an early treatment. No rafters or slatting were used, the split shakes being laid directly over the lengthwise log purlins. This style persisted into the late 1800s.

The catted chimney was often built along with the log house, to be replaced in many cases with stone or brick when the material and craftsmen became available. The log frame was lined with dried clay blocks that were replaced periodically.

building techniques. As sawmills became common, fewer craftsmen learned the finer points of hewing and shaping wood. And the use of hardware, milled doors, and windows lessened the need for close work. Probably because fewer log houses were being built as permanent dwellings by the "modern" farmers, finished work was saved for the ultimate frame house. Logs in houses became smaller in general, too, whether for lack of good timber or for ease of handling.

But there are peculiarities from region to region. Glassie tells of the uniqueness of "catted" or mud-and-stick chimneys in the Ouachita Mountains of Arkansas, an area of excellent building stone. So does Nancy McDonough. The half-dovetail notch was almost universal in log buildings of the Midwest and mid-South, whereas the V-notch was prevalent farther east. An exception is the square notch, favored in the White River Valley around Batesville, Arkansas, where many house logs were also hewn evenly on all four sides.

Beaded ceiling joists and beaded or mitered window and door trim occur more in the East. Elaborately forged ironwork is also an eastern feature.

Log house roof pitches were predominantly 45 degrees or a 12-inch drop in 12 inches (12/12), but Hutslar reports the Ohio average as a nine-inch drop in 12 inches (9/12). Some historians try to link roof pitch to age, but this, as with notching methods, is at best inconclusive. Flatter, two-story pitches of 6/12 exist in Virginia and throughout the East in all periods.

A white-pine log house from near Sugar Grove, West Virginia. The outline of the original chimney can be seen in the siding, which apparently was applied when the cabin was built in the early 1800s. The chimney had been relocated inside the walls and later collapsed.

In a given locale, builders might have favored the log gable with purlins (mostly evident in early buildings), the lengthwise poles or timbers on which roof shakes were laid. Elsewhere the gables are framed. In Kentucky, Montell and Morse found windows in the fireplace gable ends of many houses, but Roberts states in *American Folklife* that this was rare in Indiana. It was also rare in the Ozarks, except in more recent houses. It was often a case of the window glass itself being rare.

The custom of covering newly built log houses with siding seems clearly to have been practiced in areas where sided frame houses were common. It was seldom, if ever, done on the early frontier, for the obvious reason that siding was not available to the harried hunter-settlers. Where siding was later applied in the more remote areas, it is usually easy to see the weathering of logs underneath, long unsided and exposed.

Determining House Age

Perhaps a discussion of determining log house age is in order here. You will find that local inquiry unsupported by responsible records is almost worthless for houses built before the generations now living. Common responses are "It's way over one hundred years old" (200 years in the East) or "My granddad remembered that house when he was little, and it was real old then." Any building weathered gray is taken as being old, and folks have a fondness for perpetuating myths about the age of things. Let me warn you that they may also become vehement and even violent about the veracity of their pet versions of history (as any of you who have researched local precedent already know).

In general, there are a few obvious things to look for if no reliable records exist. For instance, round wire nails came into use around 1890, although cut nails can be found in houses built later. (They can still be bought today.) The cut nails generally appeared around 1820, although there were machines for cutting them earlier. Of course hand-wrought nails were used on the frontier after 1820, but (nails being a high-priority item of trade) they were rare. Cut nails are those tapered, blunt-ended ones with two rounded sides and a symmetrical oval head. Wrought nails will show the hammer marks on the irregular heads and

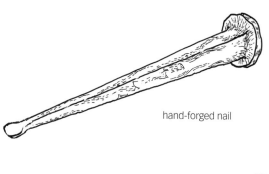

hand-forged nail

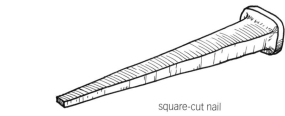

square-cut nail

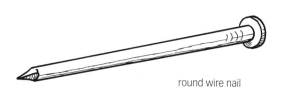

round wire nail

By the 1820s, hand-forged nails had generally been replaced by the square-cut nails. By the 1890s, the round wire nail had replaced it. Both the forged and cut nail were driven so that the chiseled point would cut across the grain to avoid splitting.

are flattened at the point, to cut across the wood's grain without splitting it.

Consequently, a log house peppered with round nails dates since 1890, unless they were added later — and their application will tell you that. Cut nails in strategic places, such as in original door facings, put your house roughly in the 1820–90 range. Pegging of rafters and window and door facings, with maybe a few forged nails, can mean you have a really historic log house — that is, if it's located in an area of early settlement. Or it can simply mean the area was remote or the builder poor.

In the 1970s, I dated a log house near my home at 1850 at the latest, because two of its four layers of roof were nailed with cut nails and the shingles were of heart cedar. This wood easily lasts 40 years, if properly applied, so even if the second roof had been put on in 1890, I shouldn't be far off. This house was also of pine logs, in an area where only the earliest houses were built of pine, because this wood is rare here now. Of course, the house could be older. But back to the roof. The one layer of shingles fastened with round nails would have deteriorated around, say, 1930, and the latest roof of corrugated iron was still serviceable, with probably 45 years of age on it. The original circle-sawn (probably since 1840) slatting over the cedar-pole rafters was also from large pine trees, and showed the marks of just those roof nails mentioned.

The use of lime mortar in chimney stones and chinking means either that a house was built before modern concrete (1880–1920, roughly) or that it could be much older, because lime and sand mortar have been used for centuries. Mud chinking could mean only that the house was remote or that its builders couldn't afford lime.

Pegging usually dates a log house before mass-produced hardware was readily available in that specific area. Again, the penury of the builder is a factor. Cast door locks became common around the 1840s, as did the pointed wood screws for attaching such things as locks and door hinges. (Before this, the screws looked as if they'd been cut off at the tips.) Look for evidence of original doors, hinges, latches, and fastenings. Many of these will have been replaced, but you will almost always be able to detect this in hinge strap mortises and perhaps old peg holes.

Many log houses were moved, or recycled, because it is much easier to disassemble and reassemble than to start from scratch. My own log house in southwest Missouri had both oak and cedar logs, hewn differently, and both half- and full-dovetail notching. I suspect the few full dovetails are evidence of reworking during an early reassembly.

We often find that log pens have been reassembled. Telltale signs are split-off dovetail or V-notches, common when handling the old wood. Or inside-out aging patinas on wall logs, indicating replacements. Or an out-of-sequence log, such as one with joist notching reused as a common wall log. Sometimes we actually find Roman numerals cut into the logs, an almost sure sign that the house was reassembled at some time.

Log gable ends instead of frame were used in early houses, with lengthwise roof poles, or purlins, for the shakes to be fastened to. I know of a few late (1870s)

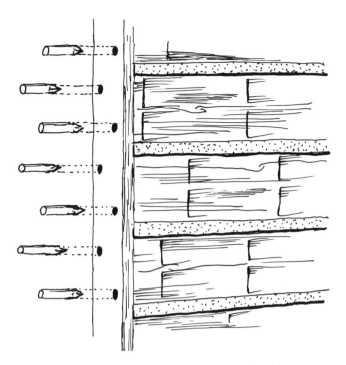

Pegged door and window facings can indicate a very early cabin, a remote location, or simply a builder who could not afford spikes. Pegs were made of locust or a similar hardwood, well seasoned to avoid shrinkage.

These "bucks" are vertical facing pieces at windows and doors, and were hand hewn. Pegs were square, driven into round holes to help hold them tight.

examples of this type, however, so the method is not sound for determining age.

Whipsawn boards almost always identify a house as being built before 1850. True, some vertical sawmills were in use afterward, but the circle saw, with its efficiency, had replaced most of them by then. Straight, uneven cuts indicate a man-powered pit saw; regular patterns mean mechanical power, usually water or steam. I know of many houses from the 1850s with rough, circle-sawn boards and beams.

Also, during the Civil War era, three-inch beams were replaced by two-inch lumber. Older houses have 3×4 rafters, where later ones will have 2×4 or 2×6 rafters and often two-inch-thick joists.

Older houses had almost no eave, whereas 20th-century versions project all around. The exception was the early house with a catted chimney, where the builder often projected the eaves on the chimney gable end only to keep the mud dry.

Study the history of the area to determine settlement dates. A claim of 1790s age in an area not settled until 1840 is obviously far-fetched. Sometimes early houses were built in isolated areas, in advance of set-tlement. An abstract of the actual property will determine the homestead date, if nothing else.

Hutslar mentions the science of dendrochronology, or comparison of log growth rings to established regional patterns. If a log's rings extend to the bark edge at any point, the exact year of its cutting can be established by matching the pattern of growth to age-known wood. Seasonal irregularities will be similar, allowing close comparison. You can assume that the original construction, or at least the hewing, took place the same year, because no log hacker in his right mind tackles seasoned timber. Check your state university for a possible growth scale.

Perhaps the most involved search for a log house's origins was that concerning Lincoln's birthplace. Weslager traces it through moves, reassemblies, rebuildings, and changes of hands to its present, largely accepted status as Abe's first home.

Whatever its age, its origin, or location, the hewn-log house has become a classic emblem of the pioneer era in the United States. It is a symbol of an age of timber, of craftsmanship, of our country's broad-shouldered youth.

CHAPTER TWO

Pioneer Building

IN THE LONG MOVE WESTWARD, the custom of building the hewn-log house came into the new territory along with the settler's hunting skills and his other crafts for survival. He might camp for months on his new land in any sort of easily constructed shelter, clearing and planting, building rail, stone, or pole fences for his stock. But his permanent dwelling — whether in the Blue Ridge Mountains, the Smokies, the Alleghenies, the Tennessee basin, the bluegrass, or on into the White River Ozarks to the west — was of hewn logs.

Tools

Certainly the earliest houses of this type in the new country were built with only the simplest tools. We know of settlers who put up houses with only an axe and no hardware. Others usually added to that tool at least one auger, and, if at all possible, nails for the shakes on the roof. Indeed, many settlers possessed or had access to the tools of blacksmithing, which made available nails and even hinges and the swinging crane for the fireplace. Plowing the new ground, replacing the axes, knives, shovels, parts of harness, horseshoes, and wagon tires all depended on a smithy somewhere within reach.

We know the split shakes of early cabins were held in place with no nails, being weighted with butting poles, weight poles, and knees, or they were wedged under saplings laid across them and bound at the ends with rawhide. But my experience with shake roofs in wind- and rainstorms leaves me no doubt that these cabins were drafty and damp.

A pioneer could and often did build his entire log house with one tool — the single-bit axe. This tool is a broad hatchet used for hewing and notching.

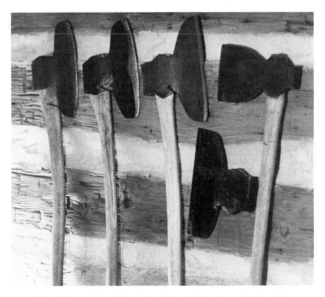

Some broadaxe types. The axe head is a 20th-century design that was still available in the 1960s. The narrow-faced axe is probably the oldest style pictured. None of the handles is authentic. The originals would have been about two thirds this length with no reverse curve.

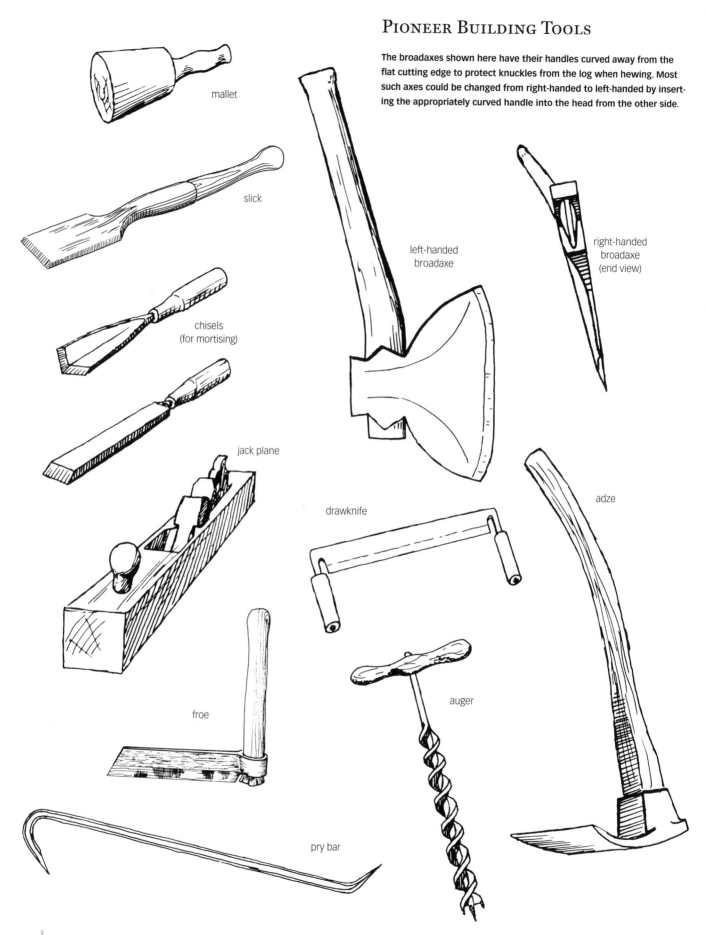

PIONEER BUILDING TOOLS

The broadaxes shown here have their handles curved away from the flat cutting edge to protect knuckles from the log when hewing. Most such axes could be changed from right-handed to left-handed by inserting the appropriately curved handle into the head from the other side.

mallet

slick

chisels
(for mortising)

jack plane

left-handed
broadaxe

right-handed
broadaxe
(end view)

adze

drawknife

froe

auger

pry bar

The building of the hewn-log house was usually done with a limited but useful set of tools, and with handmade or cut nails at least for the roof. With a felling axe, broadaxe, froe, augers, hand plane or drawknife, chisel, hammer, shovel, knife, and perhaps a saw, adze, and prybar, the pioneer could construct a house that was quite adequate and comfortable.

Site

The site chosen was usually on a rise of ground, above high water of a creek or river. Rarely was it placed in a level spot with good soil and, indeed, it appears that the site most chosen was the one least desirable for cultivation. A rocky bank above a spring was a favored spot, or rough ground between newly cleared fields. Outbuildings were scattered nearby — barn, smokehouse, corncrib, root cellar, henhouse. The hewn-log house was only one of several structures necessary to sustain the settler.

The actual construction began with cutting the trees, usually a continuing chore anyway, as fields were cleared. The man and older sons, if any, hewed the house logs on two sides while green, often adhering to the signs of the zodiac. March and May were supposedly good months to keep hewn timbers from warping. The logs were hewn to a thickness of six inches to eight inches, with the heart in the center to further reduce warping. Sometimes logs were split and each half hewn, but this was rare. Each family usually did its own hewing, over a period of perhaps months, and the logs were skidded to the house site, where they seasoned, at least partially. Then, when enough logs were ready and time from farmwork permitted, the house was raised.

The Raising

A house-raising was held if other settlers were near, and this custom was as colorful a part of the pioneer's life as we know of. Neighbors began arriving at daybreak by wagon, on horseback, or on foot, bringing their tools, the wives laden with food, with children and dogs scampering everywhere. The air of festivity was that of a social occasion, as the women caught up on their gossip, the men competed good-naturedly in

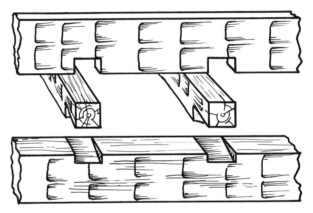

A dovetail notch being cut for a ceiling joist (see photo). Dovetailing construction stabilizes the front and back walls.

feats of strength, the girls and young men courted behind their elders' backs and the children played and shrieked over everything.

If the logs had not been hewn ahead of time, crews were put to work with broadaxes, squaring the logs, sills, joists, and beams. Then the men notched and fitted the logs, laying them on the stones of the foundation, sometimes prepared ahead of time. Notches were worked by eye, as was everything else about the early log house. Everyone knew what he was doing; they all lived in houses like this one, after all, and had done this many times before.

Sometimes the stone fireplace and chimney went up along with the logs, but these were usually added

Log gables were commonly used with lengthwise purlins. However, rafters were used here and taken out prior to moving this double pen intact one mile. It was restored at the Fresno Flats Historical Park near Oakhurst, California.

later. Sometimes the shakes were riven as the logs were laid, clay and grass were mixed for chinking, and doors were built. Rafters were laid out, and long slats split or poles cut to go on them.

The noon meal was a feast combining the efforts of all the women, and was as impressive as an outdoor church dinner. Favorite dishes combined and competed in bewildering volume, and many a stalwart lad found it difficult to return to work afterward.

When the walls were up to a ceiling height of seven feet or so, the joists were notched in. Ceiling joists were usually broadaxe-hewn on all four sides, because they were to be visible from below. Sometimes they were dressed further with the adze, and (rarely) they were even beaded with the shaping plane. Usually mortises were cut into the front and back logs through which the joists extended to the outside. Weslager shows examples of this technique in New Sweden houses from the 17th century. Wilson observes the custom of mortised ceiling joists in those log houses in Alabama with half-dovetail notches and notched joist ends in V-notched houses. Sometimes we find blind mortises cut, or those not extending all the way through. These kept out moisture, but were harder to do and generally appear on eastern houses of fine workmanship.

Above the ceiling joists, the settlers laid two, three, or even four courses of logs to give more room in the loft. The walls terminated with heavy front and back plates hewn on all four sides, which were the heaviest for the group to lift and had to go up the highest. These were pinned at each end, down through at least one set of gable-end logs. The heavy wooden pins were known as trunnels (treenails) and were whittled of tough locust, oak, or hickory, preseasoned so as not to shrink and work loose. Picture the old men seated in the shade, whittling these, talking early politics.

A final pair of end logs were sometimes notched into the ends of these plates. These logs were

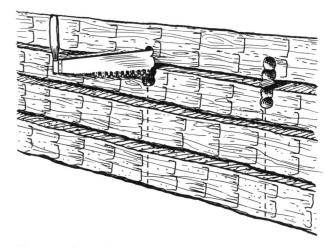

Often a row of auger holes to begin a vertical cut at a window, door, or fireplace cutout will show that the logs were assembled full length and the cuts made later.

extended, in very early nailless houses, to support the butting poles that held the knee and weight poles for the shakes. But this was only in the remotest areas, where nails were nonexistent. Depending on the building customs of these cabin-raisers, the end logs were sometimes carried on up to form the gable, becoming successively shorter to the ridge. This log gable method utilized lengthwise purlins, which were logs fitted to the ends of the gable logs. The top purlin became a ridgepole. Shakes, usually split by the settler beforehand and seasoned, were laid onto the purlins, and no rafters were used.

Still another very early method is quoted by J. Frazer Smith as involving the use of a forked pole at each gable with the ridgepole laid in the fork. Then long clapboards were laid all the way from the plate to the ridge. Weslager tells of this "crutch" roof in early Jamestown. Again, this was usually the work of a lone pioneer. When a raising was held, a more conventional, framed roof was usually built.

With the availability of hardware, the framed gable became common. It was used with rafters, and long slats were fastened across them. Now, for the first time, the sound of nails being hammered would echo through the woods.

These rafters were usually poles, flattened on top, and the slats were split or later sawn. Traditionally, when sawn lumber was used for slatting, it was not edged, but laid on full width. Spaces were left between to allow air to circulate under the shakes so they would not curl from uneven moisture. Remember, everybody knew the secrets of working wood, the signs, the kinds of trees to cut, and the use of the old tools. This was folklore in action.

Shakes were usually riven, or split with the froe, ahead of time and allowed to season before the house-raising. Shakes, like all timber, shrink when drying, so they were never nailed up green. If laid with weight poles and knees, however, the shakes could be riven at the raising and laid then. Under the weight poles, they could shrink without splitting. Traditionally, they were laid while the moon was on the increase, to avoid curling. Folk belief was that the moon, which affects tides and the rising of sap in trees, could also increase the moisture in shakes, causing them to curl in the sun's heat.

This is a Scandinavian scribed-log house in Minnesota, built in the 1890s by Finns. These logs are scribe fitted — a concave cut runs the full length of the upper log to fit the convex top of the lower log. Moss was used between the fitted logs to help seal out drafts, since no chinking was used.

This early Pennsylvania log house had corner posts with the logs mortised and pegged into them. Raising such a house is difficult, since full-length log tenons must be inserted in close sequence as the posts are drawn together.

The "crutch roof" with ridgepole used long riven boards for the final roof covering instead of split shakes. The crutch was not satisfactory, as boards warped and the roof leaked, and was done rarely.

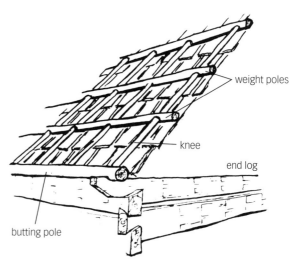

weight poles

knee

end log

butting pole

Using "butting poles" was a primitive procedure where no nails were available. Poles were often lashed in place with rawhide strips. This style of roof is and has been used all over the world by primitive peoples, or where building materials were at a premium.

The window, fireplace, and door openings were sawed out, with either a partial saw cut having been made as the logs went up or a series of auger holes bored to insert the saw through. Doors were the work of perhaps the more skilled joiners in the group, being fashioned of riven boards early on or sawn stock if a mill was near. Precious nails were used in abundance here to keep the doors from sagging. Later the angle Z-brace would appear, and on fine examples the inlet cross-batten. Wooden hinges were common, and some primitive doors were hung with leather straps. Where a forge was near, iron hinges and latch were hammered out.

The cracks were chinked late in the day, when objects weren't as likely to fall from the roof. Clay, mud, and sometimes the lime-and-sand mortar went on over split wood jammed into the spaces.

Then, as a final christening, a dance was often held at dusk, in and outside the new house. Records show that a good deal of homemade whiskey was consumed at these raisings. Indeed, Hutslar quotes an early observer as estimating that the cost of refreshments might very well equal that of paid labor, had hired hands been available.

At last the neighbors left, scattering by pine-knot flare and lantern light, to their own houses in the hollows. As social gatherings, house-raisings were welcome, festive occasions.

I have participated in several modern cabin-raisings, and these differ only in mode of transportation and dress from those of the pioneer. I hope this custom endures forever.

Building Techniques

Too often early cabin logs were laid right on the ground, with the good earth for a floor. More substantial houses were laid on large stones at the corners, sometimes set down into the ground on firm subsoil. A pier of sorts was built up to a sight level, which might be less than one foot high on flat ground or three feet at the lower end of a hillside site. Stones

A split-log or puncheon floor on a raised sill.

were laid dry, with stone chips wedged between to sta-bilize them.

Sill logs were laid, usually as the first front and back logs. Unless a log house was square, it was almost always longer down the ridgeline, so the floor joists, if any, were laid front to back (the shorter dimension) on the sills. These sills could be and were sometimes laid as end logs, but not generally. If a raised floor was to be built, joists were notched into the sills at regular two- to four-foot intervals. Sometimes small split logs, or puncheons, were notched and laid side by side upon the sills with the flat sides up to form the floor itself. These were worked with the adze when in place, and fitted to each other at the edges. As they seasoned and shrank, they could be slid together and another added to take up the space.

If dirt or puncheons in the ground were to be the floor, the log walls went up all the way to the ceiling joists without interruption. The logs were notched — either full dovetail, V-notch, square notch, or, most often in later years, half-dovetail. In rare early cases in the Midwest and East, a beautiful form of the full dovetail was used in which each log was worked with notches of the same compound angle. Rarely, too, the diamond notch, half notch, and other experiments were used.

In rare cases, corner-posting was used. The log ends were shaped to vertical tenons and inserted into mortises in heavy vertical corner beams, and pegged. This is seen more often in Canada and is associated with French building, but was done occasionally all over. The Golden Plough Tavern in York, Pennsylvania, built in 1741, is corner-posted. I also saw corner-posting in a highly crafted ruin near Goochland, Virginia — beyond repair, unfortunately. This cabin also had diamond-head king posts, the one and only example I've encountered in a log house. (I see around 100 a year.)

Logs were raised into position most often by hand. Many log houses were 20 feet long or less, or were two or more log pens of no more than that size. A man at each end of such a log, perhaps with the aid of skids, could just manage the job of raising it if no neighbors were near for a cabin-raising. Let's picture the pioneer father at one end, with perhaps his sturdy wife and a half-grown boy at the other. Or the team of oxen or

This log house has had some fanciful window treatment added, but is otherwise typical of cabins added onto as a family's needs grew.

horses could be used to cross-haul the logs up skids into place, with the final fitting done by hand.

Larger log houses, with logs up to 40 feet in length, were built where there was help to raise them. These always had midway partitions to brace the logs.

At ceiling height, usually seven to eight feet, joists were notched in, again from front to back. Upon these, as upon the floor joists, would be laid the split or rare sawn boards of the combination ceiling and upstairs floor.

Above ceiling height and more courses of logs was laid the heavy top plate, which was wider than the wall under it. In the finest tradition of log building that has survived, the rafters terminated at this wide plate to form a minimal eave. Sometimes the plate projected inward, and the rafters were notched into it or over its outer corner (the bird's mouth) to extend to form eaves.

The heavy plate also functioned to help take the outward thrust of the rafters and the roof weight. In a one- or two-story house, where there were no log courses above the ceiling joists, these joists held against this thrust, but some log knee wall above the ceiling was common.

If the house were of the log gable type, the top plate needed not be heavier. In this case no rafters were used, so there was no outward thrust.

With rafters, split slats or poles were fastened lengthwise to them for the shakes to be laid upon.

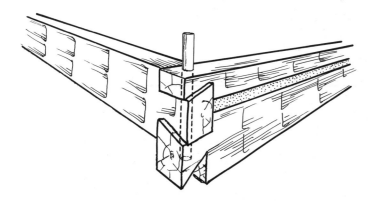

Top plate logs were pinned to hold against outward rafter thrust. Sometimes an additional gable-end log was used above the plate, as shown here.

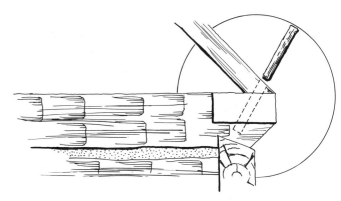

The wide top plate, supported by a longer end log, provided a minimal eave for the classic early hewn-log house. Slatting and shakes extended slightly beyond the plate.

These pieces were sometimes bound with rawhide or pegged, as were the rafters. Pegs were often square, to be driven tightly into the round auger holes.

The rafters themselves were usually poles, sometimes worked flat on top with the drawknife or adze. They were fitted together in pairs at the peak with no ridgepole. Most often a 45-degree roof pitch was used, making figuring angles much simpler. Some steeper roofs were built, and many less steep. Often the pitch was a matter of guess, or of the builder's eye — as were the very dimensions of the house itself.

Shakes for the roof were split with the froe, an L-shaped tool driven with a heavy mallet. Early shakes were quite long, often three feet or more, of the prime timber the pioneers found. If laid without nails, the shakes could be riven and used green. Nailed shakes had to be seasoned to prevent splitting at the nails as they shrank. Some early craftsmen split shakes of green wood, others of seasoned. White oak, cypress, chestnut, and cedar were most used.

If the settler had the knowledge and materials, he could forge his own roof nails from bits of worn metal. Even half a horseshoe, worn completely through at the front, could be drawn out easily to produce a handful of nails. Early blacksmiths used charcoal, where coal was not available, in a simple sand-filled forge charged with a wood-and-leather bellows. A nail-heading bar, hammer, and anvil with cutoff hardy were the only other tools necessary for this and most other simple iron working.

Occasionally, these large split shakes were pegged in place with small wooden pegs, but this was rare. Pegging was most effective as friction fastening, on the principle of a headless nail, and was not suited for the flexing of shakes in wind. The pegs would work loose, so this is where nails were used early on.

Where a log gable was not used, a gable framework of poles was pegged in, and shakes or riven clapboards were used as covering. This allowed the house to be put under roof sooner than the log gable, and the whole was lighter. As sawn lumber became available, the log gable was even less common, boards being easier than shakes or split clapboards to work and to apply.

Door and window openings, if any, were sawed with the crosscut saw. Into the cut log ends, facings of riven boards were pegged or nailed to hold the logs in line. Then shutters and doors were hung with leather, wood, or iron hinges. Shutters were of split boards, or sometimes tightly stretched animal skins. Rarely, the first settlers had glass for windows. When they did, it was usually fastened in place instead of being hung to swing open, to avoid breaking it. The wooden shutters let in light only with the outside air, and were often kept shut all winter. In hot weather they let in gnats, flies, and mosquitoes.

Doors were low affairs, made of split boards nailed or pegged together, seldom angle-braced. To keep them from sagging, the settler often used many of his precious nails in a heavy pattern, clinching them on the other side. The door was hung, and usually secured with the simple lift latch, with its latch string hanging outside.

Board floors downstairs and up were often riven, though sometimes a primitive sawmill or whipsaw, powered by two men or a water wheel, produced sawn stock. These saws were used as early as the settlement of Jamestown, and they were quite often added to water-powered gristmills in the new settlements. A board floor was a source of pride for the pioneer wife. It was pegged or nailed, with cracks lessened during installation by prying with a bar set in bored holes in the joists. Smoothing the rough lumber was sometimes accomplished by rubbing with sand and stones.

The pioneer fireplace was of stone with clay or mud mortar. The chimney was often of sticks laid like the logs of the house, lined with "cats" of mud or clay. These chimneys were firetraps, and unless used and maintained constantly, they fell apart in the rains. But they were easy to build, went up fast, and could be built with no special skills. As the mountain settlements became less remote, itinerant masons began building stone chimneys of really fine craftsmanship. Many older cabins had new chimneys of stone added after years of life with catted or loose stonework.

Chinking the early cabin was a continuing process. Short, split boards were often laid at an angle in the cracks and a mud-and-grass mixture plastered into them. Sometimes thin poles were wedged in and covered with the mud or clay. Timber being plentiful, the earlier cabins tended to be built with wide logs, leaving little space for chinking, and only later were the wide cracks of three to six inches much in evidence.

Chinking, along with the hard work of hewn-log building, has probably been the major reason for the near disappearance of log houses. The logs shrink away from the filling as they season, or anytime dry weather or heat contracts them. The resulting drafts are likely to be the most memorable aspect of cabin living to be recalled by an old-timer. Shakes, slats, boards, and, later, tar paper were nailed over the chinked cracks. Mud, clay, and lime mortar were patched and replaced season after season. According to Roberts, hewn-log houses in southern Indiana had clapboards nailed over them as soon as they were finished.

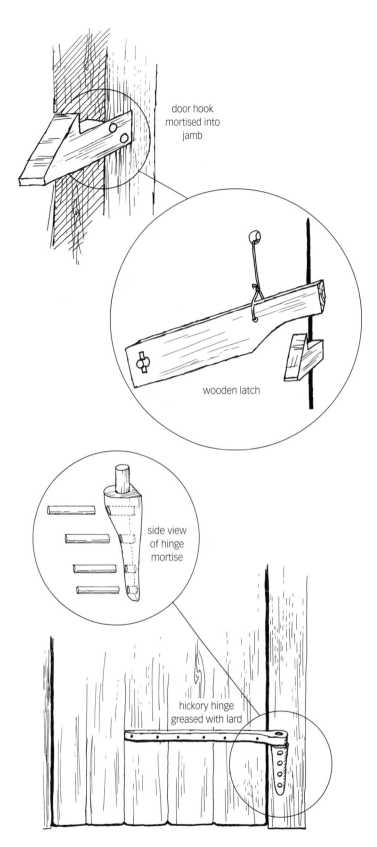

door hook mortised into jamb

wooden latch

side view of hinge mortise

hickory hinge greased with lard

Wooden hinges and latches were often used where no blacksmith was near. Made of tough wood such as hickory, the hinges were greased with lard or bear fat. The door hook was mortised into the doorjamb to allow the door to clear.

A wooden shutter on the only window opening in this log cabin. The forged hinges and latch are no longer functional, as the shutter is nailed up in an open position.

Diagonal wood splits, or chinks, were typical in early log house chinking. The spaces were plastered over with a lime-and-sand mortar.

Wallpaper, newspapers, canvas, and old sacks were applied to inner walls. Often both the inner and outer walls of log houses were covered to stop the wind, and sometimes still, a sturdy older house will surprise the remodeler or wrecker with log walls underneath.

The Pioneer Spirit

Once complete, the early log house often continued to grow with the family. A lean-to of logs or rough lumber was often added at the back. A second complete cabin was sometimes added, either on the chimney end, flush, to form the saddlebag house, or separated by a few feet, on the end away from the chimney, with roofs joined over the result in breezeway or dogtrot. A separate chimney at the opposite end completed the dogtrot house. Sometimes the saddlebag or dogtrot had a wing extending to the rear, with a roofline at right angles to the original.

Taken altogether, the hewn-log house built by the pioneers grew as a logical, practical product of the materials at hand, the environment, tools, and skills of the people. These were influenced heavily by custom and tradition. The very fact that the logs were hewn flat on at least two sides was largely social, adding to the families' community status. A round-log building was faster and easier to build, and was often used in barns and outbuildings. True, the sapwood of the round logs rotted away, and true too, round logs of the size of the hewn ones commonly used would have been unmanageably heavy. Most objectionable, however, was the tendency of wind-driven rainwater to cling to the rounded surfaces and seep its way inside.

The resulting house has survived the generations simply because it was best; the poorer efforts have disappeared because of their inadequacy. The hewn-log houses we see today are evidence of the finest of the pioneer builder's craft, weathering the centuries: proud, strong, an almost permanent feature of the hill country, beautiful even in decay.

CHAPTER THREE

Restoration

DISMANTLING AN OLD hewn-log house for restoration is a delightful unlayering of history. As the original family grew or the house changed hands, the tastes and circumstances of the owners were imposed on the aging structure.

Taking off lath and plaster, removing crumbling additions and porches, step by step getting back through the generations, finally we reach logs. And we sense the comings and goings, the toil, and the sudden joys of so many who made this their home.

The inside of a window casing holds a penciled record of a visit by one girl to another in October 1838. Bees have built hives in the hollow walls of this addition in the 40 years since it was lived in. In a pocket of debris is a newspaper scrap with the chilling ad for "a well-mannered carpenter for sale." It is so hard to relate to life as it must have been lived then.

We never find money. Old bottles beneath floors, lavender with age. A lathed-over window, complete, in what had been an outside wall. A row of coat pegs in a basement, with a sliding wall panel to more room in an addition — this was once a tavern.

Once, we found a newspaper from October 1859 that included an account of John Brown's raid on Harpers Ferry as seen by an eyewitness. The paper is stained and faded, preserved between furring strips and lathing over a log wall. So that's about when this cabin had its face lifted — when the nation stumbled toward its bath in blood.

These owners wanted a respectable house. False graining decorates those widened stairs. Three rooms have elaborate fireplace mantels and hearths, fake against the plaster walls of additions.

The Virginia farmhouse pictured here (inset) was built in 1859–60 around a late 1700s cabin (above). We researched the house, stripped it, and restored it to as near the original as possible. It remains one of my favorite restorations.

A No. 1 Carpenter for Sale.

WE have for private sale a likely NEGRO MAN years of age, who is a No. 1 Carpenter of good ral character, and sold for no fault. A₁

PULLIAM & CO., Odd   Hall

oct 18—ts   Fra   stre

RICHARDSON

This newspaper ad was found in an 1838 timber-frame addition in Goochland County, Virginia.

Jack posts hold up this log house while a slab basement and footing are poured. Later a timber frame–and-stone wall was built here.

These pine logs with brick nogging between them were covered with siding over 200 years ago when the house was first built. The logs showed no weathering. The original doorway to the left of the center log partition had been converted to a window sometime in the past.

These nicked logs once held a stucco covering on a historic log house. Furring strip marks also indicate that it was sided over at one time with clapboarding. Log houses were often camouflaged.

Life went on, as politics and war and commerce and the westward movement continued. The logs, the houses, hold in time a little of the settler, the farmer, the pioneer: We get a glimpse of our shadowed past, a hint for understanding our present.

We moved a tiny chestnut cabin recently, redone and added to and under coats of white paint. It was next to and probably predated a huge Victorian shell that had been cut up into apartments. The cabin, we learned, had been a rural Virginia doctor's office; the house was his home. What stories these two could have told. Both are gone now, and the hill slope carved flat under the asphalt of a crossroads shopping center. Only the cabin lives, rebuilt a few miles away on a hillside over a lake.

The Challenge of Restoration

Restoration is sometimes so very involved. A caved-in basement has required us to hold up a house on jack posts while we dug out with a backhoe. Leaking basement walls have resulted in our digging completely around the house to below masonry level, sometimes to install footings a little at a time. Then we've sealed the walls, drain-piped, graveled, and back-filled.

We've redirected the flow of underground springs that had undermined foundations. We've propped up a house to replace logs cut into blocks for indiscriminate window placement. We have stabilized bulging walls that had no reason to stand any longer.

I enjoy the challenge of re-creating a cabin from the barest of evidence. Penny McGee, one of our favorite Virginia clients, showed me a caved-in stone basement and the stub of a chimney on the slope of the family's Cherry Mountain land. A larger foundation joined this, and this frame house was what the older neighbors remembered on the site. A trace of chestnut sill log was still in place, with part of a dovetail notch. The dimensions, 18 by 28 feet, were clear, but not much else.

This is a case where the restorer has to fill in the blanks with what he knows and can find out about the area. Steep land can mean later settlement, but in this case there was a farmhouse from the 1700s just a mile away. The chimney and basement were dry-laid stone, whereas other area stone was lime-and-sand mortared. But that could be as much a factor of penury as of age.

We found a chestnut cabin five miles away that was vacated in 1871, the date the present farmhouse was built. This cabin was one-and-a-half stories, with double-hung six-over-six windows, and a stone chimney. And it was 18 by 28 feet. So that was actually a replacement, not a restoration. Really, it was a restoration of the number two cabin on the number one site.

Some challenges I don't want. For example, there was a Depression-era, second-growth pine clubhouse thrown up on a lake, with no foundation at all — the logs on the ground uphill and on loose stones downhill. Over the years, subsequent owners had dug out the rotten and termite-infested lower logs and pushed in mortar. They'd propped up the sagging floor joists on stones and bricks, and termites had a highway up to wood. They'd built flower boxes and a flagstone patio up against the log walls that held moisture against the wood. But it looked charming in its setting of big beeches at the end of an inviting road. It ranks as the worst job we ever undertook.

Another repair or restoration challenge I decline is any work on a log house kit. Repairs to sapwood pole houses are another breed of cat. I have even been asked to be an expert witness in suits against kit manufacturers and franchise builders to assess the caliber of the material used for these modern trips into nostalgia. Enough said.

Warning

The first thing I do for an owner contemplating restoration is try to determine just how much he or she really wants this house restored. It must be among the most vital, driving needs for it to be worth the money and effort.

There is no way to anticipate all the damage in a house until everything is stripped away and the house stands blushing in its pristine nudity. More often than not, a restoration requires that the house be dismantled completely, or at least stripped to bare logs in order to get to the roots of decay.

It is often cheaper to build new than to do a major repair job on a log — or any other — house. That is why we so often build new structures using antique

Old flooring is often an asset in restoring or recycling a historic structure and requires careful removal in order to reuse the material. Removing heart-pine tongue-in-groove flooring from a house prior to dismantling it poses challenges to safety. I have one crew member on each project whose job it is to back nails out of reusable lumber right away so people won't step on rusty nails. These log joists, flattened on top, are typical in early cabins.

logs, beams, paneling, and doors. That way everything is bug-free and works — plumbing, heating, all the modern stuff that makes living in the house better than a chore. But everyone thinks, after about a year, that this is a historic house.

I don't invent histories for my new houses made of old materials, but I suspect some of their owners do. It's hard for eyewitnesses to its building to disprove the origins of a "200-year-old" house.

You must also know that dismantling a complete house — no matter how small it looks — is no fast, safe, or easy chore. It is hard, hot work. Stepping on rusty nails becomes a given. It will — using my favorite formula — take three times longer than you planned. And it will cost at least three times more than you budgeted by the time you find the house, buy it, hire a crew or bring in your friends to take it down, find vehicles numerous enough and large enough to haul those long logs and beams to your distant site, unload the house pieces, and find a place to store them until you are ready to set it up again. Then begins the adventure and cost and time and effort of the setup. You get the idea. You'd better be serious, because you are about to begin a real-life experience.

Beginning

Both building and restoring the hewn-log house have much in common. But all in all, there are compelling reasons for restoration instead of building from scratch.

There's the difficulty in obtaining materials and a justified concern for dwindling forests. Certainly you will save a lot of time and labor using seasoned, fitted logs instead of laboriously shaping them yourself. Whether the house is to be restored at its present site or moved to another, you're often three months or more ahead in acquiring materials.

But that isn't the reason I like to restore old log houses. Perhaps I can best tell it this way: A good friend and I were dismantling a long-abandoned house I'd found for him, which he planned to move and rebuild. In the attic, still dry under its replacement tin roof, was the cast-off accumulation of over a century of living — old farm catalogs, a quilt pattern collection cut from old newspapers, cinema schedules, shredded clothes, ladies magazines. Faded scraps of letters, gnawed by mice, told bits of history. We were able to trace the growing of a lad named John, born just after the turn of the century. Some school records, letters reporting his staying with relatives out of state — fragments. We thought it quite fitting that my friend's son, then six, would also grow within these old walls.

People come from everywhere to watch a log house being built, rebuilt, or torn down. The young ones are usually just curious, but the older ones almost always have a store of log house memories of their own to relate. The old man who shuffled up as we removed the last of the roof was one of these.

"This house used to belong to my folks," he said. "I grew up here."

"Your name isn't John, is it?"

"It sure is. How'd you know?" He was genuinely glad we were going to rebuild the house again, and promised to visit later. For my part, I considered it something of a privilege to have been a part of the restoration.

I have sketched bits of family history from old cemetery stones, and tried to match the births and deaths with additions and changes in the nearby houses that still stand. A lean-to may have been added, for instance, about the time of the birth of twins now in their 80s. Or the building of the second half of a dogtrot may coincide with a settler's marrying a widow with children.

Different broadaxe strokes also bear the tale of patient craftsmanship or of hurried shelter construction, as do marks of other tools, details of foundations, care in stonework. Sometimes there will even be penciled notes on walls. I recall a house in the Buffalo River country of Arkansas that has a regular diary written on the outside wall, dry under the porch.

Of course the best reason for restoring a log house is to preserve a bit more of our vanishing heritage, reflecting so well the American pioneer culture that has allowed us to grow, for better or worse, into what we are today. If you believe in vibrations, they are certainly abundant in the silvering logs and old hearthstones laid so long ago.

Recycling material means a lot of handling, hauling, and storage.

Finding Your Restoration Project

Finding a restorable log house is first. Then buying it or, less likely today, getting it for the act of removal and cleaning up the site. Then there's dismantling, pulling lots of nails and removing the inevitable additions of years, then transportation. Then you're about ready to begin at square one.

You'd be surprised at how many ways you can find a log house. In the mid-South, there are adventurous souls who find, dismantle, and advertise log houses in newspapers, in magazines, and, of course, on the Internet. A number of professionals are listed, at this writing. Start with a search for "log cabins" and you'll be surprised at what you'll find.

Often if you buy the house through your restoration contractor, you will be better served. The contractor will offer a warranty on the house purchased, and will pay greater attention to detail and to satisfying you, the client. The long-distance supplier of the log house will find it difficult to respond to many problems, while your contractor will be working with you every day. The supplier will probably offer to replace decayed material, but my experience has been that I have a better chance of getting the supplier to make good on the logs and other materials than the individual clients have. First, I know what is good and what is not; and second, the supplier knows it and knows that I will not accept poor-quality materials. So when you pay the contractor, you are paying for a guarantee of the right material and the best-quality material.

Of course, there are other ways to find a log house. Country newspapers are a source. On a slow day, the editor of a weekly will often run a feature on somebody's grandfather's house; maybe it's for sale. Or a farmer might advertise logs for sale. Or you might run an ad yourself seeking logs or log houses. You might even have a log house in your family without realizing it. Ask around. Keeping the house in the family might appeal to that aging uncle who was using it to store hay.

Locate your house by asking, driving around, or tracking down newspaper photographs. I will say that cabin hunting cannot be a used-car-lot pursuit. It takes time, and you should never roar up to a backwoodsman in your shiny vehicle and grill him impatiently about log houses or anything else.

This West Virginia log house shows the condition of many cabins you will find. This cabin had not been lived in for many years, yet had logs and other materials in excellent condition.

Wide logs such as those in this house from near White Sulphur Springs, West Virginia, mean less chinking and a more efficient restoration. This restored house is near Earlysville, Virginia.

My friend Bill Cameron, then almost 80 (and for whom my brother and I helped restore Turnback Mill near Halltown, Missouri), was the finest hand at discovering antiques I've known. Be it old millstones, log houses, lost cemeteries, or sorghum mills, I seldom knew him to fail. It went like this — we approached a farmhouse or country store in the requisite battered pickup truck, and he visited with whomever was there. In maybe five minutes he'd found that some old friend or relative was a mutual acquaintance (sometimes I wonder if he invented these people), and reminiscences followed. We were soon invited to join in anything from dinner to a hunt for just what we came after, or we were referred to someone else who could supply it. I suppose the best introduction to country folks who know about log houses is, finally, being obviously country folks yourself — not hobbyists — with calluses to prove it.

Once found, it may take a lot of persuasion for the owners to part with the structure in question. They may have plans for it, but get them to thinking about selling it, and that will work on their minds. Check back often. An example is a man I worked with in Memphis who'd been after his cousin in Pulaski County, Tennessee, to sell him one of several cabins and log barns on the extensive old family farm. The cousin had plans to restore them all. Then a heart attack made it clear he would never get to these projects in whatever remained of his life. He called the Memphis cousin and told him to come take his pick. The result was a chestnut and poplar cabin with 28-inch-wide hewn logs, moved and restored as a guesthouse. Those were among the widest logs I've ever seen, although I frequently work with 20- to 24-inch ones.

Surprises

I caution again of the work and danger involved in this project. Start looking for the red flags right away:

- termite damage
- water damage and rot
- logs beyond repair
- caved-in basement or foundation
- loose chimney stones or bricks
- disintegrating windows or doors
- sagging floors
- decayed joists
- bowed rafters
- decayed roofing or roof decking
- worn-through floors

The first enemy of your restoration is the very work of time that may appeal to you most. Unless your find has been boarded over and roofed with tin, the materials can be pretty crumbly. A log house can look sturdy as it stands but come to pieces when dismantled. Most alarming is the way bark and rotted sapwood flake off, leaving you with six-inch chinking cracks where the originals were two inches.

Pine and poplar rot from the outside, and logs like these may have massive, sound heartwood. But oak and chestnut usually start decaying from inside, when water gets into check cracks. These logs look better than they are. Often a thin shell hides a spongy, rotted core.

Check your prospect for soundness. If a log looks rotten or termite infested, it probably is. Held in place by the comfortable stresses of years, logs and beams can look solid when they're not. Poke, pry, hammer on, and stick your knife into everything you can reach until you're satisfied. I recall how solid an unroofed chestnut log pen in Virginia looked. Most of the logs were mush inside.

A house that has no roof left is a risk. The old, dry logs go to pieces very quickly when they are exposed to weather. As little as three or four years uncovered

When notches, such as this full-dovetail one, are split off, it is sometimes necessary to reshape them and use a wood block for a spacer. This one was later trimmed off flush to fit.

can reduce some log walls to rubble. And if some of the walls have caved in, forget it. Once apart, the logs will generally be worthless.

I once looked into restoring a cabin near Roanoke, Virginia, that had been dismantled and left on the ground for four years. Almost none of the logs was usable. The beaded, hand-planed joists had been banded together and had rotted each other. I did not do that restoration.

Look for houses with wide logs. That means more heartwood and less chance for rot. The chinking may be gone or intact, but chances are you'll lose some sapwood either way. Pole cabins of less than eight-inch logs usually aren't worth the effort.

Buying

Acquiring a log house you've found can be harrowing. Often the least desirable hulk becomes suddenly treasured beyond price (almost) when you evidence interest. Having established a good down-home relationship with the folks helps. So does the fact, if made visibly obvious, that your resources are modest (which is why you want that old cabin, anyway — to save the cost of a new house). I have acquired log houses free, just ahead of a developer's bulldozer. I have paid more than they were worth for desirable specimens. And I have just plain talked my way into possession.

Perhaps the most involved experience I can relate was in Searcy County, Arkansas. The cabin I wanted was just about roofless, but had a wealth of history, some recorded on century-old tombstones in a nearby cemetery. Inquiry through a friend who was related to just about everybody in the county revealed that two brothers owned this cabin, along with other abandoned farmhouses. But these were only two of eight heirs to the holdings, scattered from there to California. Neither brother would sell without everyone's permission.

A Fourth of July reunion was scheduled for that summer, and I got the brothers to ask the others. Meanwhile, the cabin was moldering. No help for it, though. I was off to college that summer to try and finish an elusive degree, and heard no more till late August.

Seems when the kinfolks realized which cabin I wanted, they had all said yes, and so I tried to strike a price with first one brother, then the other. Neither wanted to say, and each referred me to the other. Finally my friend and I set out to dismantle the cabin, taking along most of a jug of relatively good homemade whiskey given to us by some local candidate for office. (They still did things like that in the early 1960s, after you got to be known as good old boys.)

You see, one of the brothers liked a nip now and then, so this was the one we went to see. Well, he helped with the jug some, and finally allowed that I could just have the cabin, because he'd probably only use the logs for fence posts, if that, even. And he helped us take it apart. True story. I wrote an article about that cabin for the September 1969 issue of *Ozarks Mountaineer* (a regional magazine), under a pen name.

If you buy a log house from an absentee owner, get a letter authorizing you to remove it. Muskrat Murphy bought his Arkansas cabin (which she planned to move to Missouri) from a doctor in Michigan, and her letter kept us out of trouble with zealous neighbor folks who weren't right sure we should be taking it apart.

Don't even fondly imagine you can slip in and whisk away an abandoned log house undetected. First of all, it takes a lot longer than you think, and everyone in the country will know about it. Some students of mine and I once four-wheeled down a nonroad miles beyond the reach of most folks to dismantle a one-room cabin in two hours flat. But in the time we were there, we encountered an entire horseback riding club, a hunter, some canoeists, and a farmer on foot.

Cost

A final word on buying your cabin. It will be quite expensive to dismantle and transport it to your site. So unless you buy the land and restore it on the spot, don't pay much for the house. I was offered a borderline condition dogtrot log house for $1,600 in 1976, 200 miles from home. All told, I'd have had probably $25,000 in a barely livable restoration before I was through. So I said no thank you. If it had been free, I might have taken it, more to preserve it than as a bargain.

Dismantled, you're talking about nearly 100 logs up to 20 feet long, in this example, plus perhaps good

We had to unscramble these miscoded logs to get them back in order. The Dodge Power Wagon with winch and gin poles is our log-lifting equipment.

This dismantling shows how dusty and dirty old houses can be. We let down these logs on Muskrat Murphy's house with ropes to keep from breaking them.

rafters and possibly rare, whipsawn roof slats and handworked beams. That's several truckloads, for a long truck. If you pay anything at all for labor, you'll have more than you'd like to admit tied up in just getting the logs to your site.

A complete house might provide you with other material — flooring, doors, joists, paneling, wainscoting, fireplace mantels, rock, brick, window and door trim. You might find that you don't want to reuse the flooring as flooring — for example, the boards might become window trim. Wall boards might become cabinetry wood. We once found 29-inch-wide boards used in a stairwell pantry that became a gaming table for my wife. Chestnut roof decking became our kitchen cabinets. So yes, the material might be worth a lot, but don't put a lot of money into it, then a lot of labor, too.

Buy cheap. How cheap is up to you, but whenever building, always multiply a reasonable estimated sum by three to arrive at what you'll probably spend.

Once bought, you have two options in moving your house. A house mover will move it intact, but it costs a lot of money. The Grigsby dogtrot log house on the Arkansas College campus in Batesville was moved 12 miles intact except for the chimneys. Cost was around $10,000 in the mid-1970s. That's not unreasonable for a 50-foot building. In 1988, a similar house in central Virginia was moved two miles through fields and woods for $80,000.

Dismantling and Coding

The usual method is to completely dismantle the house. Location-code each log while the house still stands, using a carpenter's crayon or indelible marker so rain won't wash off your code. Mark in the notches or chinking surfaces so the code won't show. The third log up on the east side could have a 3EN (north) on one end and 3ES (south) on the other. You can also do this with chimney stones, joists, and rafters. Where you have windows and doors (and therefore many short, spacer logs), the code can get involved: a 3ENL (for left), a C for center, an R for right, for instance, or subdesignations like ABC where applicable.

My worst experience with coding came in 1990 with the purchase of a fine, wide-log heart-pine house from near White Sulphur Springs, West Virginia. The supplier had stood at the front door and used a north, south, east, and west coding. But he'd mixed east and west, so the notches didn't match. If I had not done business with this man many times before and trusted his skill, I'd have caught the mistake sooner.

Often I'm called in to unscramble a pile of logs with no coding. Even a photograph helps here, or at least some memory of where the front was, or the chimney. Lacking any hint, I lay out the logs, exploded fashion, on the ground, using the obvious outer weathered surfaces as the first guideline. Sills, if any exist, will be obvious, as will top plates and the notched spanners that held the overhead joists. The fireplace opening will be wider than doors or windows, and is almost always in the gable end, or shorter dimension.

Short spacer logs, which were left between windows or doors, are hard to locate. Remember that most early cabins were built of complete, long logs with the openings sawed out later. So plot the course of each sectioned log from big notched end to small notched end, by size, taper, and signs like matching checking cracks.

Dismantling is slow and dangerous. Heavy logs, sometimes with nails protruding, are prone to fall on the unwary. And more than once, I have encountered snakes reposing in hollows where chinking had fallen out. Bees and buzzards love abandoned buildings and make their removal an added challenge. Rats, dogs, and birds have left their spoor (and desiccated carcasses) everywhere, and dust is usually thick. Once I even removed a roof to let the rain wash down the rest of the cabin, but it didn't help much.

Respirators are necessary for everyone on the project. Get good ones, with replaceable filters. They'll cost $20 to $30 apiece but are well worth it.

My favorite dismantling time is winter, when poison ivy, snakes, ticks, chiggers, and wasps are dormant. Winter landscapes are better for spotting cabins, too.

Perhaps more involved than most dismantlings was the Murphys'. It was a community affair, sort of a house-razing, if you will. We converged on the impossibly high mountain in Newton County, Arkansas, with an assortment of four-wheel drives, one heavy pickup truck, and several trailers. It was a weekend in May, just after a lot of rain, so several folks got stuck in mud holes in the old logging road that wound up the hill. It was here we were challenged by a neighbor who informed us the road was closed to the public, and who never really believed the letter we had from the former owner.

I spent most of an hour trying to convince him we were all good old boys, not being helped at all by the sunglasses and colorful garb of the others, while they got their vehicles unstuck. Finally he and I recalled a mutual bulldozer operator acquaintance, who was then in jail, he informed me, and talked about some other folks I thought I dimly remembered. He finally even offered to help us out of the mud, except that his back was sort of out, from having "drove fence posts yesterday," but by then we were ready to go, anyway.

Well, it rained all weekend. I was more or less gainfully employed at the time, and was expected to be on hand for a special event at work, so my wife, Linda, and I visited the site late each afternoon. We witnessed the soggy camp as it grew soggier, the demolition of the cabin and of the foodstuffs, the dampening spirits.

By Sunday we decided to evacuate the pickup and its tandem-wheeled trailer before the alternate road became impassable, too. This road ran straight down the mountain over boulders higher than the red clay was deep, so of course we high-centered often. Then we winched back up to bring down most of the short

Restored cabins showing log and frame additions.

drystone pier

Building codes will generally not allow loose stone pier foundations. A continuous foundation must have vents and metal flashing between the masonry and logs.

logs on a trailer behind my Land Rover. On the way back up I slid into a ditch and overturned, which delighted the owner of a Japanese four-wheel drive, until he tried unsuccessfully to pull me upright.

That's chapter one. A couple of weekends later, Muskrat got together another set of friends and rented a huge truck to complete the move. But no one would drive it up that gullied road, so I volunteered. Never have I been so banged about as I was inside that metal cab. The steering wheel cracked me around like a whip, but we got to the top of the hill.

Then, after we'd loaded many thousands of pounds of logs on the beast, everyone sort of looked at me again. Well, so much for a misspent youth driving log trucks. In I climbed and down the mountain we lurched. That truck was never the same after that.

The story would not be complete without the house-raising that followed some months later. A third set of friends (notice how all but a couple of them seem to have learned from experience?) gathered on one of those rare sunny November days to raise the cabin.

There was lots of good food, hard work, camaraderie, and even suspense (Murphy had coded more than one log with a given code number). We almost flattened two of our number under a plunging 400-pound log from high on one wall. The kids and dogs loved it all. Later there was some fine mountain music around the campfire.

More Cautions

There have been other near misses. Once my brother John and I tossed a top-plate log onto a pile of split shingles, nails, and other roof debris below, reasoning that this would cushion the log's fall. It did that, along with my own, when a protruding nail caught under my wedding ring and I was jerked off the high wall. I came down with the log, somehow staying out from under it, to smash onto the scrap pile, bristling with sharp points. Miraculously, none of these pierced me. The worst pain was a sore ring finger. My wife understood when I stopped wearing the ring. In another incident, a friend and I were taking down a log barn in Nelson County, Virginia, and started an avalanche of roof rafters when we removed the vital pair. We dodged the big 4×5 timbers nimbly, choosing the rare patches of upstairs flooring that hadn't rotted through.

These and other episodes should encourage you to be extremely careful, or to hire someone else to do the dismantling. I have multiple old injuries now, which prove that I wasn't always so lucky.

A candidate for restoration often becomes nothing more than a pile of logs. Roof framing is usually brittle or rotten or, if sound, is too lightweight for reuse. Slender poles or sawn 3×4s were common as rafters, which held the shake roof well enough then, but now don't allow for insulation or inside loft covering.

Building Codes

Building codes come into play here, too, which in most instances require minimum 2×10 rafters and nine inches of fiberglass for R-30 insulation. The codes also specify ceiling joists that are usually heavier than those in old buildings. This can mean you'll have to find other, preferably also old, material for these uses. The small stuff can be used as stair rails or furniture or sawn into boards for window, door, and wall trim and other uses if it's in good shape. We reuse 100 percent of the sound heart pine that was common in eastern log house construction, but not often in the same ways it was originally used.

Much of your log house may not meet the more or less standard U.S. one- and two-family dwelling codes. Stair pitch and width, ceiling and door height, specifics regarding foundations, floors, and framing may fall short of today's requirements. The best course is to visit your county building inspections office with drawings and details of your project. Where it is impractical to make changes, the inspector may grant you a variance. In a recent case of mine, there was a problem with door height. The six-foot eight-inch standard would have meant cutting through the spanner log above, which carried all the ceiling joists in their mortises, so I was given a variance for a lower doorway. Sometimes steeper stairs will be allowed, too, if there's no room for the less-than-45-degree code requirement (nine-inch minimum tread; eight-inch maximum rise, at this writing).

If yours is a significantly historic house, you may want to preserve it exactly as it is, without modifications to code. If you explain this to the authorities, along with the fact that this house is for you, not to be sold to an unsuspecting buyer, your chances of receiving the necessary variances are better. Building codes are for the protection of us all against shoddy workmanship (as well as to add revenue to the county budget), so they have some basis in reason.

Features you can easily change in a log house that you dismantle and move (as opposed to an on-site restoration) you will be expected to modify. If the house sat on dry-laid cornerstones, you'll be expected to rebuild it on a continuous foundation of stone, brick, concrete blocks, or poured concrete, properly vented and on the required inspected footings. You can, of course, veneer over block or concrete with stone or brick, or plaster over (parge) either one to cut down on the uglies.

How Usable Is the Wood?

Keep in mind that all of your wood may not be usable. Sound, original heart-pine tongue-in-groove flooring is, and will likely remain, in high demand, and is worth removing carefully, numbering so as to get it back in order (wear patterns make this necessary). Often, though, over a basement or near the ground, you'll find powder-post beetle damage, little pencil lead–size holes that lead to big eaten-out caves in the wood. Unlike termites, which have to be able to reach ground moisture, these insects can and will reach any wood that's "punky" (soft) enough to be lunch for them.

Make sure the wood you reuse is sound, not only for strength, but also to avoid infesting your restoration with voracious little eaters. They don't do much damage to heart oak or pitch pine heart, or to chestnut, which is virtually all heart. But sapwood pine, poplar, oak — just about any wood that's been wet long enough to get punky — is not worth reusing.

This, of course, applies to the logs themselves. We usually cut away the sapwood at the top and bottom of old logs where it was often left on with the bark. This gives us a log hewn on all four sides and a sharp lower corner to recess chinking under so it'll stay dry. It also means a wider crack to chink than originally. But this is no problem because modern chinking with a wire lath will allow this. Sometimes this sapwood is

Often a restoration requires the entire existing house to be rechinked inside and out. Before rechinking this Virginia log house, all the narrow cracks had to be widened by removing sapwood and bark. Taking out the punky wood is good for the health of the house as it leaves only solid wood in place.

removed when the logs are down and sometimes it is chipped away in the standing house.

Often old logs will have decayed sapwood, top and bottom, where chinking has held moisture against them. If the decay runs all the way across most of the face, we cut away a section of the log face and apply a new face of similar wood, then rechink. If decay is all the way through to the inside, we may splice in a similar log section. In that case, we set a block in under the splice to brace it, set back so chinking covers it.

A common decay area is at a check crack, where water has gotten in and rotted the crack. Here, we cut out a neat rectangle as deep as is necessary to find good wood. Then we cut a matching piece to fit, glue it in, and set it with galvanized nails. We caulk to keep rain out.

Rot at the log ends is also common if water has been allowed to get into the notch. Replacing a notch is trickier since the notch functions just as much to hold the log in place as it does to secure the corner of the house. If too much of the wood is rotten, we will lift the log end out and replace a section of at least two feet, making a new corner notch at one end and a splice into the existing log at the other end. If the rot can be contained in the center of the log end, then we remove the rot, make a plug, and insert the plug into the log. We worked on one house that required a plug more than two feet long. The outside of the log was solid, but it was hollow from rot. One of our carpenters, who is very tall, with very long arms, was able to

reach into the log end all the way up to his armpit to remove the punky wood. We had fun with that one.

Wherever decayed wood is cut away, use a heavy application of high-quality wood preservative to kill dry rot. We use a borate solution, brushed or sprayed on at least four times. Several brands are on the market, none of which is toxic like the older fungicides.

Here is a good place to point out that if you're replacing or recycling a log, always use the best and widest face out — the one with a good bottom corner, at least — to tuck the chinking under. Inside, the log face can be almost round, because there's no weather to keep out.

Log floor joists are often too warped or too weak for reuse. In this case, we replaced only the decayed ones.

Floor joists, often of logs flattened only on top to carry the flooring, will sometimes be warped too much for reuse. I will usually replace these with 2×10 saw-milled stock when it won't show and if there's to be no basement. If it is to be visible from below, here's where I use newly cut beams, often cut down from recycled old factory or warehouse timbers. Lacking these, you can use fresh wood if it's been seasoned for at least a year. Or you can buy kiln-dried heavy beams through most building supply stores for a price.

The rule of thumb for joists is one inch in height for every two feet of span. That's misleading, though, because a 6×10 can carry more weight than a 2×10. And although modern building codes require a basic 16-inch-on-center (o.c.) spacing, you may be replacing beams set three feet apart, so they'll have to be bigger. Also, what you're flooring them with affects the size and spacing. In our house in Virginia, we have three-inch grooved, splined flooring from an old cotton mill; this spreads the load considerably, so we have fewer but larger (6×10) joists.

Here's where you need that conference with your building inspector. If your log house had 2×6 ceiling joists on 24-inch centers, for a span of 16 feet, it may have held a century of upstairs living, but he'll want something more solid. Given the rule of thumb, you should have 2×8s that are 16 inches o.c. or heavier, but you won't want to look up at all that zebra-stripe framing above. So suggest 4×10s or 6×8s, maybe 36 inches o.c., or more, to fit into mortises in the wall logs where the originals had been. That'll mean flooring with two-inch-thick tongue-in-groove (t.i.g.) upstairs, but that's a good idea anyway. You can then perhaps reuse the old, one-inch-thick flooring from up there over a sub-floor, downstairs where it'll be seen and appreciated.

We've gone from roofs to floors here, so let's go up there again. You may want to keep those pit-sawn 3×4 rafters, especially if they're mortised and pegged nicely at the peak. You can use them as a base for your roof, decking with boards; you may paint between the rafters to lighten your loft. Then you can do a built-up roof, using 2×6s or 2×8s on edge, toenailed on top of the decking, with insulation between and the roof covering on top of that. It's expensive, but it looks nice.

We usually rebuild a roof using 2×10 rafters for more insulation and to meet the building code, deck-

These hewn 6x10 joists in our stone kitchen addition are floored upstairs with three-inch-thick splined cotton mill flooring.

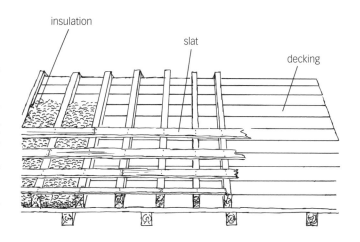

This drawing shows a double or built-up roof, which lets historic heavy roof rafter beams show inside while fully insulating the roof itself. These beams are strong enough to hold up the roof but too shallow to be seen and still accommodate the thickness of insulation. This complicated roof is a trade-off between the ambience of seeing these vintage beams and spending the money and time.

ing and covering conventionally with shakes or standing-seam metal. Then we insulate between rafters and wall the insides with shiplapped boards or drywall, often painting off-white to lighten the upstairs. This gives a flat plane, not as interesting as the old rafters but a lot cheaper. To add interest we'll often reuse the old collar ties, or new ones made from old material, out in the open up above head height. These can be left with the old saw marks or planed and oiled for the beauty of the wood grain.

Patching Logs

When encountering rot in a log — in either an existing log wall or a log to be installed — you first remove all rot. Each problem requires a slightly different approach, but if you don't chop out every bit of the rot, then you're only building in problems for the house. Next, paint a borate solution onto the old log. Then, make a patch and fit it carefully in such a way that it will match the wood. All patches need to fit into place with glue and caulk, and set with galvanized nails.

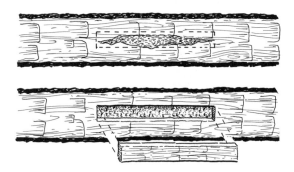

Common decay at check crack in log. Cut out section to get out all decay. Replace with patch of appropriate wood.

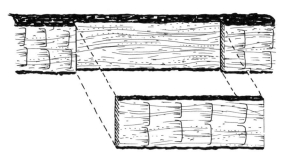

Sapwood is often decayed into the face of the log. Remove bad section as deep as necessary and reface with matching log piece.

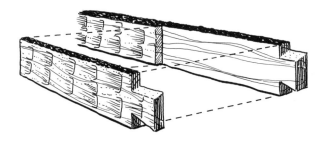

Cut out mortise and fit in matching piece.

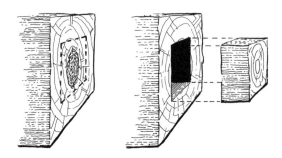

Remove punky wood from the center of the log. Cut out a square space and make a matching plug. Paint hole with borate solution and glue plug in place.

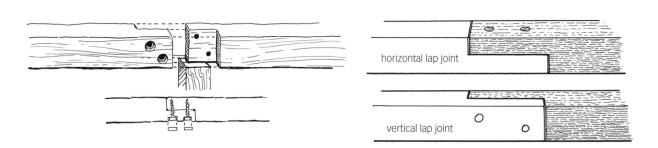

horizontal lap joint

vertical lap joint

The half-lap joint for splicing replacement log sections is perhaps the simplest way to repair a long log with limited damage or to join two log sections to create a long log. This joint can be made vertically or horizontally. The horizontal joint hides the bolt heads/pegs better. A vertical line in the log is neater looking. However, the choice is mostly personal preference or based on the condition of the wood. Use lag bolts to attach the joint, but recess the heads and plug the holes with wood to hide the bolt heads. When wooden pegs were used instead of bolts, no attempt was made to hide the peg. A lap joint made in a log wall should be blocked up under the split; the recessed block can be hidden in the chinking.

In general, even on log house restorations, a roof is a roof, and you may choose your treatment. Because heat rises and is lost mostly through the roof, this is an important place to heighten energy efficiency while maintaining as best you can the old look and craftsmanship.

Here's a hint: If you want to show rafters and collar ties but can't reuse or get good old replacements, you can have new ones sawn and stained to darken them. The circle-saw marks from modern sawmills may kill the effect for you, however. When I want as authentic a look as I can get from new wood, I either hew it by hand with broadaxe or adze, or I have it bandsawed. This can be done with one of those popular portable units. Bandsawing leaves straight marks similar in appearance to those of the old pit-sawn pattern, which are much more pleasing. Of course if you plane the wood, it doesn't matter.

Engineering Feats

If you're restoring your log house on-site, you will of course want to keep everything that looks sound. Sometimes even restoring a log house on-site requires major if not total dismantling. Unfortunately, it's often necessary to dismantle in order to get at or even to see all the damage. But if it is not major, some replacements and repairs can be done piecemeal. Some of this work is relatively easy. Replacing major logs, such as sills or spanner logs, becomes an engineering feat.

Most often there will be decayed logs to replace. And most often these will be at the bottom in the hardest place to reach. Settling foundation stones or dirt washing against the structure can put the sills in contact with the ground, where rot and termites can go to work.

I have seen termites eat up through a corner notch necessitating the replacement of logs in both directions. If they have chewed only the sill or any logs on one log face, you can support the whole wall. Carefully lift out the bad log or log segment and replace with an age- and size-appropriate log.

To replace lower logs, we support the next log above using a piece of channel iron about two feet long with hydraulic jacks under both ends, one inside and one outside the house. Then, instead of jacking

Lap-jointing a log into a short corner to extend and stabilize the log wall, which had been cut away long ago. Every second log was replaced entirely for stability. The wood will be wire-brushed to blend better.

The matching half-lap is cut in the replacement log. When working near other homes, we often use an electric chain saw to cut down on noise and fumes.

up the house, which can crack windows and chinking, we take just enough weight off the bad log to cut it into pieces and get it out, corner notch and all if necessary.

Then we install the replacement, whether sill or wall log, of similar, aged material. Obviously the corner notch will be a problem, because the log end won't go through the smaller space left for it. We split off the end, drive the new log through the opening, then use construction adhesive to glue the split piece back in place, clamping overnight.

If the corner notch is good, it's easier to half-lap a replacement section onto the original. We lap at least

Decayed wood has been removed here, and the borate-based preservative is being applied to stop dry rot and possible insect damage. We use four coats, brushed on or sprayed on.

Construction adhesive is applied to the replacement log section which, in this case, was most of the face of this log.

The replacement section is put in place. Finish nails, along with the glue, will hold it. Clear caulk is used to seal out water.

12 inches, using a vertical lap instead of horizontal so the patch shows less. It's a good idea to leave the replacement piece a little long, say half an inch, to allow for adjusting the butt ends of the lap joint. If they don't fit tight, kerf the gaps with a handsaw to let them come together tighter. Then, when the joint is tight, glue, peg, or nail, and cut off what's left at the end, if anything.

The lap joint can be done without jacking up the house, so of course it's preferable if the notch area is sound. All too often, though, the notch is where the damage is, and it must be replaced. If only the notches are bad, you can do the same half-lap repair, replacing the notched section and using the split-off technique mentioned before.

If the house has settled, which it probably has in this case, do raise it in small increments over several days, back to just above level. Drive replacements (logs, stones, wedges — whatever) in tight, then let the house wall back down. The logs will compress until rechinking distributes the weight from above along their entire lengths.

For higher replacements, we use jack posts, which adjust to most workable heights. One goes inside, again, and one outside, with channel iron between. Avoid automotive-type jacks under vertical posts, because this arrangement can buckle and let the whole house come down — sometimes onto you.

Rechinking

This is a case for a very close look at the condition of the log house before you start, or better, before you part with any money. Rechinking is just about always necessary. If anything turns off the enthusiasm to restore a log house, the dilemma of replacing rotten, unworkable chinking is it. At best, the expansion of the logs with moisture has pushed the old chinking out, to leave a direct channel for rainwater behind the masonry. Often, the logs have rotted from this, or the old chinking is so broken up it has to come out anyway.

I have found the chinking space between the logs filled with solid clay and solid concrete as well as wedged with large rocks, small stones, full bricks and broken bricks, moss, grass, mud mixed with horsehair or straw, and wood splits (or "chinks," the derivation for the word *chinking*) smeared with a layer of mud or concrete in varying thicknesses. (More recently, chunks of Styrofoam have been wedged in.) Early builders did not have access to metal lath and the materials we have and so they stuffed anything and everything they could into the cracks. By doing this, they actually sabotaged the health of the logs, and therefore the house, by holding moisture against the wood.

The best thing to do with old chinking is to knock it all out. Then, if the log wood is soft or bug-eaten,

ADDING A COURSE OF LOGS

This cabin's first-floor ceiling height was oppressively short (just over six feet). The ceiling was raised by adding a course of logs at the top of the door and window height. These photos show raising a cabin wall with the roof in place.

UPPER LEFT The upper section with roof in place is raised with 20-ton jacks to open space for the added log course. Replacements and repairs have already been made to the end wall.

LOWER LEFT With one log replaced, the upper section is raised further to allow space for the new end log. Both the heavy jack and the one jack post are used to keep the full-length log above from breaking under the weight. Temporary props keep the loose sections of log in place. Note where the window buck stops; that is where the log course was added. You can see how much higher the ceiling was raised.

BELOW Here I'm setting the new end log in place, with help from inside. The sliding brace at right lets the upper part of the cabin be raised but keeps it from shifting to the side. One end of the 30-foot cabin was raised at a time.

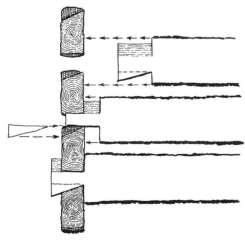

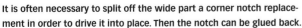

It is often necessary to split off the wide part a corner notch replacement in order to drive it into place. Then the notch can be glued back.

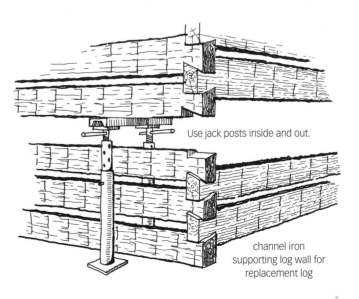

Use jack posts inside and out.

channel iron supporting log wall for replacement log

We raised the corner of a standing log pen to splice in a replacement log. In this case we used a finger-joint splice on the log being raised with a ratchet hoist. The upper section is lifted here with gin poles. If the roof is still on, use heavier house jacks to lift the roof and log walls. This is really not an exercise for the amateur, but the result will make living in the house more comfortable.

The finger-joint splice going together. The joint was glued with construction adhesive and pegged in place.

cut it back to sound heartwood, even though this will mean a wider space to chink. Be especially careful to cut back horizontally at the bottom of the logs, to leave a corner that the chinking can be sloped up under so rain can't get in.

Once the bad wood is removed, cut and nail in the wire lath just as you do in new construction. It is sometimes necessary to cut away the whole face of the log to get to good wood — and sometimes, of course, you never reach it. That's when you know it's time to splice in a replacement. We've often had to cover repaired chinking with siding when the necessary surgery is just too ugly to bear.

In one case, we were called in to rechink a lakeside cabin whose weathered face had felt the brunt for many years. Not only had the solid chinking pulled away from the logs, permitting water to enter and rot the logs, but recent repairs had only aggravated the problems. Styrofoam had been stuffed into the cracks and cemented over. By the time we axed away rotted wood and filled the chink crack with 12-inch-wide chinking, the logs certainly looked sickly. We conferred with the owners and the unanimous decision was to clapboard that side of the house. We managed to spare more log face on the inside, where it retained the look of its log heritage. (For more about chinking, turn to chapter 11.)

Plumbing and Electricity

Besides chinking and log replacement, you'll almost certainly need to replace or install electrical wiring and plumbing. Once you come up through the floor with the electrical circuits, the wire can be run in the chink spaces. We attach electrical boxes either horizontally or vertically; often we mortise the boxes into the log faces themselves. This takes longer, but is much neater.

At doorways you'll need a way to get up to switch height, and also to go on upstairs. If possible, it's best to remove old chinking and facings from the logs at the doorways, then drill vertical holes through the logs. You can sometimes pull one log inward, to drill it and the ones below and above at one time. This can be difficult, however, when long logs are involved. If you are re-erecting the logs, that drilling can be done easily in the exposed logs. If the logs are already in place, an angle bit will accomplish the same thing with a bit more difficulty. Above door height, the spanner log can usually be drilled to let the wires go on to upper levels. We use a one-inch bit.

An alternative is boxing the wires into grooves behind the door trim, although this makes them vulnerable to nail punctures. Putting the wire into conduit will help.

Plumbing is always a problem in log houses. Pipes from above can be enclosed in chases, hidden in closets or in kitchen cabinets. Older houses did not have closets, so the chases become good excuses to build them in.

But upstairs plumbing must have traps, and this often means boxing in part of a beamed ceiling. When laying out upstairs bathrooms, it's a good idea to keep the main drain near a downstairs wall, and not to cut across and weaken main joists with it.

Vent pipes must be installed within set distances of traps according to building codes as well. Finding ways up and out for them can be a chore. You may well have to settle for location of fixtures in strange places to ensure that they'll function. We went up a shower wall in our house with the vent pipe, then out through a log and up only a short distance to get above eave height. Two stories up, it's not that noticeable. (For more about plumbing and electricity, see chapter 15.)

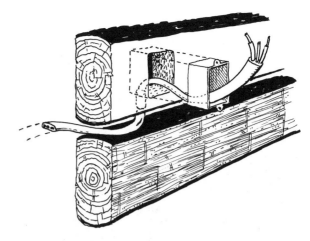

Electric wiring can run horizontally inside a chinking joint and vertically in a trough at window and door openings. Do not place the box in the chinking joint. The best way to install the electric box for outlets, switches, and fixtures is to mortise the box within the face of the log. This is more difficult than shortcuts like surface-mounted boxes and chink joint-mounted boxes. However, this gives a more professional-looking installation and allows you to place these boxes exactly where you want them.

Decayed wood or a too-tight space requires widening the chinking crack for new work. A hatchet is one of the best tools to do this.

Windows and Doors

Restoration of windows and doors is tedious work, and gets expensive. Often doors have sagged and have been cut off-square to keep from dragging on floors that have sagged too. Many old cabin doors are too narrow or quite short (often six feet or less) and must be replaced to avoid head-on collisions with the overhead log or the building codes. Stretching doors with patching always looks sloppy, although we do it when there's no alternative. That's the case with National Historic Register work, which we document. When you do it, it's a patch — no excuses. Just do it neatly.

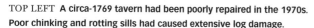

TOP LEFT **A circa-1769 tavern had been poorly repaired in the 1970s. Poor chinking and rotting sills had caused extensive log damage.**

TOP RIGHT **The tavern being dismantled. Earlier work had included pine sills close to the ground, which were decayed after 25 years. Also, the granite chimney stones, set with Portland cement, had to be reset using lime-content mortar.**

BOTTOM LEFT **The restoration nearing completion at its new site in Virginia. Replacement material was documented.**

BOTTOM RIGHT **The completed restoration overlooks a lake and horse farm.**

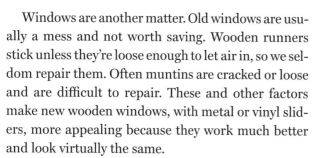

Windows are another matter. Old windows are usually a mess and not worth saving. Wooden runners stick unless they're loose enough to let air in, so we seldom repair them. Often muntins are cracked or loose and are difficult to repair. These and other factors make new wooden windows, with metal or vinyl sliders, more appealing because they work much better and look virtually the same.

We often save old wavy glass and put it into the new units, maybe one or two panes per sash, to keep the old look. We have installed old glass in hand-built windows and ordered commercial windows without glass. In those houses where the owners were sticklers for authenticity, the new windows with the old glass added a special touch to the house.

Restoring the originals is sometimes in order, as in Historic Register houses. Then we buy, or have machined, shaper bits to cut replacement stiles and muntins. We use stable wood such as fir, clear ponderosa pine, or often close-grained heart yellow pine for the replacements. Usually the jambs are also worn, so we rebuild them.

Dollarwise, a carpenter can spend two or more days restoring a window, which is easily twice the cost of a new one. But that's the essence of restoration; it has to be worth it.

Compensations

You may not be able to find a suitable house for restoration within reason or in the right place. Most of the following chapters are devoted to skills needed both to restore a house and to build your own step by step. If you restore, these construction details and skills can make the difference between a good restoration and a disaster.

If you need floor information, refer to that chapter. Roofs always need replacing, so thumb the pages of the roof chapter. I've arranged the book this way partially for this purpose.

There are compensations for the rigors of log house restoration — lots of them. Building your own house from scratch or restoring one will be lots of work and take lots of time or money, or both, so you have a choice. If you build new, you have the tremendous satisfaction of hewing your home from the forest and laying the clean, bright wood together to start its new life as your shelter. If you restore, you have the sense of history, good vibrations, and the knowledge that you have saved an important part of our past, and some standing timber too.

RESTORATION CHECKLIST

- ☐ Look for danger signs — rot, damage, etc. — before purchasing a structure to restore.

- ☐ Look for houses with wide logs. More heartwood means less chance of rot.

- ☐ Plan on spending much more than the cost of the house itself for the complete restoration.

- ☐ Do the dismantling slowly and carefully to avoid injury.

- ☐ Code logs meticulously. Your job will be much harder if codes are confusing or wrong.

- ☐ Maintain a good working relationship with your building inspector.

- ☐ Plan to remove all old chinking and rechink.

- ☐ Add plumbing and electricity to comply with building codes.

CHAPTER FOUR
Land and Site

WHITE-CANVAS-TOPPED WAGONS wound west into the new country 150 years ago, up the rivers and the little lost creeks, past laurel thickets and into the beechwood glades.

The settlers stopped their tired teams beside clear pools and springs. They camped and looked around them — at the black soil, at the mountains rolling to the sky. And sometimes, when the mists had blown out of the hollows next morning, they stayed. And built.

A clearing for a garden, with the inevitable stand of corn, was first in order. Shelter was often an overhanging bluff for many months. Sometimes it was a tarpaulin stretched from the wagon bows to the ground. For days and weeks, the sound of axes echoed up the spring creeks, and threads of wood smoke rose from the new campfires.

Logs were burned on the spot or rolled aside to be hewn for the house. Then, sometimes a year or more later, the settler and his wife and older children raised the cabin, or if neighbors were near, a community raising was held. Hewn on two sides using the broadaxe, the logs were carried or dragged to the chosen site.

Water was the first consideration for that site, and early cabins were located near springs or streams. A favored location was against a rise of hill, overlooking the sloping floor of a hollow that was to become fields, down to a creek or river. If a side branch or spring ran by, even better. These were also the sites favored by the Indians for their hunting camps and villages, and many a pioneer homestead was littered with shards of pottery and bits of stone bowls. I have found deeply hollowed stone mortars in old stone fences throughout the mountains.

The evening damp brought chills, and settlers tried to build on higher ground, even if it meant carrying water. But the blufftops, far from bottomland fields and exposed to winter blasts, were also avoided. Only after successive generations pushed their claims up the ridges were the rocky tops cleared and homesteaded. As the first settlers' children and then their children grew and spread back up the mountains, sites became more remote, less ideal.

Many of these ridgetop cabins still survive — tumbled ruins left stranded when the automobile came, unable to follow the wagon roads up the hollows.

This Ozarks homestead is located on river-bottom fields of the Little Buffalo, which runs against the bluff beyond.

A steep site is not necessarily a bad one. This Virginia location is deeply shaded in summer but affords a view of the lake and fields year round.

Finding Your Land

Today the same requisites for a cabin site apply as they did 150 years ago. Small acreage is generally hard to find. The expense involved in dividing a large holding, surveying, and providing abstracts for every piece (the record of all transactions in the past) makes reasonably priced land sell only in large units. Tracts of 160 acres or more are easier to find, and usually much cheaper per acre.

If you look long enough, you will find a good tract, for a price. There may be an old foundation there, or perhaps jonquils in the spring gone wild from some settler's path no longer visible. For enough money you can still find a prime location.

And that leads us to the inevitable question: How *much* money?

Land is worth whatever you think it is. Owning an acre of ground, to do with it what you wish, is worth a lot to some folks. And small parcels cost more per acre (lots more, depending on location). As part of a large, more remote holding, the relative cost of a few acres

diminishes. For insulation against the rest of us, you should have five acres or so, on which there may be one or several acceptable house sites. Unfortunately, everyone wants a five-acre plot, and that puts the price up. We'll get to relative prices later.

Consider a large tract and find friends to buy the rest of it, keeping your choice plot. Or consider ruining your bank account for larger acreage, with the idea of selling off some later. Or buy it anyway, sell your second car, and give up all your other expensive habits. Land and a log house will take all your energies anyway.

Land purchase is a one-time outlay, unless you later mortgage it to pay for the kids' college. You can find out current prices from those deceptively attractive real estate catalogs, from local real estate agents, and from just asking around. But in the end, this one big expenditure is your choice, and is worth it or not to you alone.

The right land must be so right you just can't let it get away. Then price becomes less awesome. You get this feeling of belonging to this one hill, or wanting to dig your hands into its soil and become a part of it, and you know this is the place. If you don't, you probably shouldn't even entertain the prospect of log building.

So the land doesn't have a spring. It could be full of phosphate anyway. Maybe that wet-weather waterfall runs only four months out of the year. Invite friends only on wet weekends. But that white oak tree is 200 years old; Daniel Boone's contemporaries scouted around it.

You can always drill a well, and maybe push in a road without ruining the whole mountain. You can get the local cooperative to extend the electric line, or use kerosene lamps (avoid most schemes for generating your own). And you've searched for a year, and argued with your spouse, and you're tired.

So buy it. If you can live with it, whom do you have to justify it to? Well, the banker, mother-in-law, the guys at work . . . Nonsense. This place has the right vibrations. It's your personal statement. Here is where you build your log house.

Keep looking for the site that is so right you can grow to belong to it. The right spot is worth a lot of inconveniences.

Cautions

These are a couple of things to watch for. Whether or not you buy through a real estate agent, insist on clearing up anything questionable on the abstract. Title insurance is often substituted for an abstract. Read all fine print. That option to drill for oil, cut timber, or other evidence of mineral rights forgone by previous owners could still be active. If there's any doubt about property lines, have the land surveyed by a licensed surveyor. It costs, but your building site should not be too close to the line. Who knows what may become of the adjoining land? A hog farm? A chemical plant?

Watch for liens; make sure you have good, legal access; don't take any rights for granted. The romance of opening three pasture gates to get to your haven pales in nocturnal sleet storms. And a new owner next door may have other ideas about the route he will provide you access on. If you buy with others, get it all down in writing. When George and Sally split up, she may end up with it all and sell it to a paper mill. On this note, it's better to own your part free and clear, and keep a friendly relationship with the others. Ideals wear thin when it's mortgage-payment time.

I might mention that my wife, Linda, and I found 200 acres bordering the National Forest, which we decided we must find a way to own — that is, at least part of it. After much balancing of eggs, cajolery, drawing and redrawing of imaginary division lines, we were able to get a brother and three close friends to buy parts of it, leaving us with our choice of some 40 acres of clear creek, waterfalls, and bluffs. We go there whenever we can and swim in the creek and listen to the silence. Someday I'll build a log cabin there, very near the site of a 100-plus-year-old settler's home.

That's one way the land purchase can be done in a group, and there are certainly others.

Think Ahead

But don't create a place so weird or remote or with such limited appeal that you're stuck with a turkey if you do need to sell someday. This ended up happening to some friends of ours. They built a delightful house up from a long, clear pool in the creek, down a long, very difficult limestone ledge road. The house

This log boathouse under construction is on a lake in north Georgia. It is part of a major new private building project using antique logs and stone.

was a jewel, but far beyond electricity and the reach of a well-driller.

They reasoned that the inconveniences would be minor compared with the beauty of the place. Drinking water came from a neighbor's well. They used a generator to run machinery and rechargeable batteries to run the stereo and computer. They pumped creek water to a rooftop tank for utility water. A wood cookstove helped heat the house, and both the refrigerator and summer-use, two-burner stove were propane. They were very organized, meticulous, and everything went like clockwork.

And they had several very good years there, while inflation raced on. Sure, they both had to find work away from their homestead from time to time, but for the most part it was good.

Then their goals shifted, as goals will. About the time they turned 30, they realized there was a lot more world out there to be experienced, and the shortcomings of their place (and of each other, incidentally) became evident. Everything they had was tied up in the house, land, barns, and garden. So leaving meant selling.

Like most of us, they hoped to find special people whose likes and tastes would fit the place. Unfortunately, also like most of us, they found these special people are usually also poor. No one with the money to buy would consider their 1½-mile ledge-rock road and creek ford, or their electric-powerless, non-running-water house.

With a good $100,000 (much of it in labor) tied up in a unique handcrafted log house with stained-glass windows and full gas kitchen; horse barn and pig operation and numerous sheds; fruit orchard and well-cared-for garden, these people ultimately had to just about give away the place to someone else who'd be willing to spend most of his time subsisting. The inconveniences translated here simply and clearly to dollars lost.

A dream gone bad. Even if this couple stayed together and on this land, their needs, values, and aims had changed; this situation would have grown worse for them. They now look at the adventure as seven good years, on which they've each closed the doors with few regrets.

I know, however, that $50,000 or so apiece from the place would have made it a better experience to look back on. Now it will cost much more to do something like that again, and the capital just isn't there.

I know lots more stories like that. Our own Missouri place, though equipped with heated floor, air-conditioning, dishwasher, two fireplaces, deck, two

GREAT LOCATIONS

TOP LEFT **This site atop a mountain in western Virginia has sweeping views, but is exposed to weather on the north and west.**

UPPER RIGHT **Restored cabin on the site of an old schoolhouse near Gratz, Kentucky.**

ABOVE LEFT **The cabin replaced a modern frame house that was to the left. It is above a spring-fed pond near Boonesville, Virginia. Of oak logs, with copper roof, stone and cedar gable siding with no paint, this cabin was designed and built to be as low-maintenance as possible.**

LOWER RIGHT **Upper porch view of old fields and woods above the Hardware River in central Virginia. This is a new cabin of white pine logs.**

bathrooms, and lots of other goodies, was hard to sell. The county road was more than two miles of gravel and steep. My driveway was 400 steep feet. The site was five steep acres. The house was a restoration and looked its 150 years.

We finally found the right buyers, after a lot of money spent advertising, and then a 10 percent commission paid to a good real estate agent. But we didn't find that mythical buyer with cash, hungering for our place. We financed it over several years.

Choosing the Right Site

Enough about selling, which is the last thing you're thinking about at this point. In deciding on a location, you're probably pretty much mobile; most of us are today. And you have probably chosen your general location for reasons peculiar to you — a job, family, climate, schools, beauty, and so forth.

Your specific property should supply your specific wants and needs. On or near a good road. Near schools. Forested. Open fields. Great view. Privacy. Close to neighbors. Far from neighbors.

Water. Even if you never garden, there are few things as soothing as a spring creek or a waterfall. In the ledge rock of the Ozarks, ours dried up come summer. If you have visions of turning a grist mill or generator with a water wheel or turbine, better have streams with good flow on your prospective property. And watch for flooding. My miller friends get inundated and have to muck out every spring.

The springs have been polluted most places I've lived. Lime-rock country lets the water follow underground channels without filtering, which can sometimes give you a distant city's sewage. You even get soapsuds in your well if the geology is wrong. Sandy soil helps, and so does a remote location, with no septic tanks near.

In my work with log house building and restoring, I've seen some pretty sick people who, for instance, haven't seen the need for letting the bird droppings wash off the roof before the water goes into the cistern.

Compromise, sometimes. When you really get down to the specific location of your country place, it's going to depend on what's available. Holding out for the last requirement on your list of 19 will mean letting some good places go by. And you get older in the meantime.

You'll also find yourself shifting priorities as you look at land. It's like buying a new car; you start revising your needs to fit whatever turns you on at the moment. That's called salesmanship by the folks who sell cars, and land. Later you, the buyer, wonder how it all happened.

Friends of ours lived on a mountaintop in West Virginia, miles from power and roads and schools. They homeschooled their five children, worked at whatever odd carpentry they could find, and sold handcrafted woodwork at craft fairs. Most of their time was spent subsisting — hauling water, gathering and feeding wood into the cookstove and heating stove, keeping lamps lit, caring for animals, and making their garden productive. And they're not sure it was worth it. The children have had some trouble fitting back into society.

I'm not saying you should go looking for land with an inviolate list of requirements, with eyes and ears closed to more sensitive appeals. I'm saying that with some major considerations allowed for, the right place, when found, will be really right. Infatuation can get expensive, in both dollars and years invested.

Forty years ago, I let a place on the Buffalo River bowl me over with its beauty, to the exclusion of just about all the practical considerations. Now my family and I have a much more reasonable place, with its own charm to fit our admittedly changed affinities. And it's certain to wear better.

My idea of a real estate salesperson is one who shows the land and lets you and it work out your relationship alone. You've heard it said that agents work for the seller, who pays them. (Legally they do.) This is not true of a good one. Good agents know the dollars come from the buyer, who's going to be in their area, starting now. They work for both, in that demanding game of matching people and property. And, if they're good, you'll see an interest beyond that of handling a salable commodity, beyond that of the money itself. When a salesperson cannot effectively hide the boredom that comes when property all gets to looking alike, I look for someone else.

You can contract with real estate agent for them to act as "buyer's agents," employed by you, the buyer, to secure property for you. They are also bound by law to disclose all a place's warts to you ahead of time. Often it's worth it.

I'll sum up the business of locating your land by saying you should establish some clear guidelines, with an eye to the future. Realize that you and your situation will change, and allow for that change. Then let the magic of the right piece of land work within those guidelines. It will, and you'll know it.

And about how to buy it? Any damn way you can.

Buying It Anyway

Once you've found it and fallen in love with the land, the money for it must be found. I've always proceeded on the assumption that if I really must have a thing, a way to get it will appear. This seems to work, too. I remember I had to borrow most of my part of the down payment for the Buffalo River place. Then my brother and I tried every financing arrangement we could think of, finally locating, through the real estate agent, an individual who'd loan the money as an investment. I've found money in strange places. And when it just can't be done, maybe that wasn't the right purchase anyway.

I won't go into a lot of detail about price, because that's entirely up to you. It's got to be worth it to you alone. But do compare, so you don't become a victim of greed. See what the other comparable tracts are going for to know where to start figuring from.

Here in central Virginia, rural land goes from $1,000 an acre (much more if close in or in small acreage) to $30,000 or so per acre. Up in the mountains, past electricity and roads, you can do some serious bargaining. Recently an acquaintance bought 200 acres of mountain for $600 per acre.

Land prices today are outrageous. But I can't think of any time in history when that wasn't true. Even in depression times of very low dollars, those same dollars represented a lot of human endeavor. So the two-dollar-an-acre land we hear of from our grandparents was really out of reach, even then.

If you go way out in the mountains for cheap land, you'll probably keep on paying for it in road construction, maintenance, vehicle and travel cost, and quality of life.

Convenient land is higher in price, but you generally pay for it only once. Taxes will rise if the county is prosperous, but there are often benefits from that, too. Good roads, schools, a sane growth policy, culture, all often come with higher taxes. Look at your priorities.

Legalities

You usually buy land if not for cash, then by deed and mortgage, or by contract for deed. The first gets the place secured in your name, with a loan against it. The second means you get a deed when it's paid for, at which time it's transferred to your name. You pay taxes, insurance, and so on either way. Contract for deed makes it easier for sellers to repossess if you default in payment, because it's still sort of theirs on paper. Most states have now begun to recognize that an equity built up by this method must be acknowledged in case of repossession.

I've bought land both ways. I've also taken and given second mortgages in which, say, you as the buyer pay the sellers their equity in payments (with perhaps a down payment) and mortgage the land to a bank or savings and loan association for the balance. It means making two payments. Also, the sellers must be willing, because your default gives the bank first consideration in case of repossession.

It's nice to have the cash. It's nice, if you don't, to have the seller finance the place. Individuals will usually give you a better interest rate than lending institutions, presumably because they don't have the overhead in employees, computers, and flamboyant buildings.

Camp on Your Land

When you've found several good pieces of land, go back again and spend some time there. I like to camp there if possible, see each property in as many lights as I can. Compare. If you can't make up your mind, chances are you haven't found the right place yet. You usually know it when you're in love, even if it comes as a surprise. And if you're drawn to a place but can't quite say why, you'll probably keep on making delightful discoveries after you've bought it. That's serendipity.

We camped on our present house site in the fall of 1978 during a 6,000-mile odyssey in our Land Rover. A mountain rise to the north, and we noticed the wind went over our heads while the sun bathed the southwest slope. We pitched our tent where our living room is now, and spent some time getting to know the place. It has not disappointed us.

Building Sites

Look for good building sites, or maybe a restorable building on that good site. A good site will have some of these features:

- exposure to the sun along the house's long dimension
- accessibility sufficient for a short or minimal maintenance driveway
- site not too steep, so hillside will not have to be carved out and soil held with retaining walls
- soil among rocks
- view, if possible
- shelter from prevailing winter winds
- water available, even if deep drilling is necessary
- soil that will allow septic percolation

A family I know is moving from an enchanting six acres and fine hand-built house after six years of fighting their north slope, with its lingering snow, steep road, and icy blasts all winter. I like a south-facing, sheltered site, maybe just down from the brow of a hill.

I used to snort at the building guides that said build on flat ground. No way was I about to locate on a level just to save a few dollars on foundation costs. But washed-out driveways in constant need of repair and maintenance go along with a slope. This also means general erosion, the necessity of climbing to or from the house to your car, and some other ongoing problems (a high foundation can mean freezing pipes under there).

So maybe you don't want a flat site, but one that's within reason. Rainwater used to cascade down the steep hill onto our Missouri house and surf up against the logs, to come in, along with mud and sand — despite our caulking. To stop this, we filled, stone walled, and rerouted the drive up above, which was an ongoing chore of some two to three years.

We held this Missouri hillside in place with the stone retaining wall and steps. Erosion is always an issue on a steep site.

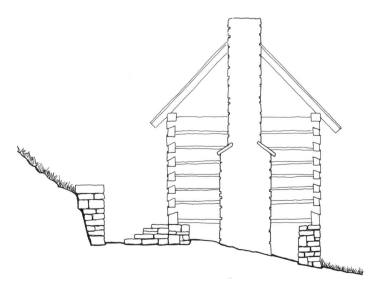

A well-built retaining wall should slope into the hill it's supporting.

A livable house on your property is an asset, providing shelter while you build. Or it can be a source of material for your log house.

People have always hung houses on ledges, dug them into slopes, perched them on spaces just foundation-sized. (This seems to be a fetish in California.) This is almost always a mistake, as time goes by. A mudslide or earthquake tells you why.

So locate a good building site before you part with your money. Remember, you'll drive or walk up to your house thousands of times in the years to come, and it must be in the right place, look right, and, in short, the house must belong there, like it sprang up after a spring rain.

Land with a House on It

Of course if there's already a house of sorts on the land, it's probably in the wrong place, or facing the wrong way. We found lots of solid, restorable houses with the character we wanted, but every one of them was either too close to the road, aimed wrong, or just on the wrong part of the land. Too many times houses were put on the worst piece of ground, where crops or livestock wouldn't grow.

If there is a restorable building there, you need to know something about the costs of reclaiming it to figure into the purchase price. Charm comes with a high price sometimes, and it often includes termites. Unless the house is livable for you at that moment, it's best to figure it a liability — with no dollar value. And tell the owner or real estate agent so in no uncertain terms.

The Site

All right, the land is yours. You have several sites in mind. Find the spot you want, the view you want, the approach to your house you want, and unless it's physically impossible, build there. (If it's level, you'll never pass for hill folks. They have one leg shorter than the other from walking around hillsides.)

The first thing you will "build" is the driveway to your site. That driveway becomes a very important aspect of your life and lifestyle. Build an approach to your site that is both attractive and practical. A good driveway will make construction easier and cheaper. And it will create fewer maintenance problems and feel right every time you drive up to your house.

Points about the house site itself: Can a well-driller get to it? Or is it close enough for you to pipe water? Those rigs are enormous, and their drivers seldom respect a graceful limb that's in their way or want to risk their equipment on a pitching road. Will you have to carry materials in? Logs are heavy, and don't bend well around tight turnings of your path. Is there room to maneuver a small trailer, or to get a pickup truck in (a real one, not a toy)?

Is there good, porous soil for sewage drainage and filtration? The lime-rock ledges of the Ozarks and parts of Tennessee and Kentucky and part of the Shenandoah Valley are terrible for effluent filtration. So you'd have to make plans and budget for a more efficient sewage system.

Does the land slope too much? I've built five-foot stone walls on the downhill side to perch a 16-foot-square cabin on, but I didn't like it. We built seven feet up on the west slope of Greene Mountain in Albemarle County, but I wouldn't want to go any higher. You wouldn't either.

The best way to handle a steep house site is to build a higher downhill foundation, instead of cutting down into the uphill side. You get mud, clay, and erosion by digging and leveling. We did both because we had a flagstone floor and had to level. It was a cut-and-fill operation, done by hand, but it took years to heal the scars. You can build a retaining wall of stone uphill to hold that dirt in place.

How about shade in summer and exposure to the wind in winter? A ridgetop will give you power for a windmill on your well and maybe a nice view, but it will frost you out, too, in winter. Being too close to a creek or wash can leave you soaked, or leave you with your foundation washed away. Watch for loose rock up steep mountainsides, too.

And access is also important. We finally gave up trying to drive the last few yards to our Missouri log house and built a stone retaining wall up the slope. We filled in a turnaround area above, with steps down. It wasn't too far to walk with groceries or babes in arms, and those last vertical feet had been hard on clutches and tires. Besides, we'd rather have the mossy rocks and plants than a vehicle and muddy ruts all around the door.

Site Preparation

Whatever you do, don't turn loose a bulldozer operator to build you a road and level your site without your supervision. Be there every minute, and yell loud. I've operated dozers, and can't explain the power thing, but it's there, and it's hell on trees and earth. It's your

Try not to destroy the landscape with the road to your site. Above all, the road should look as if it belongs where it is. Visualize the approach to your house; you will make that approach many times. This driveway required only minor brush clearing.

land. These are your plans. With care, a dozer can be made to work for you. If you must level, do so sparingly, remembering that torn soil washes away badly. Use silt fences to help keep runoff in place.

My most frustrating experience was near Ponca, Arkansas, when I'd located a delightful site to nestle a cabin for friends in a grove of five beeches up a slope from the Buffalo River (before it was made into a national park). The dozer operator was good, following my route up a natural bench, hiding the road neatly all the first day of work. But I was held up the next morning. When I arrived, I discovered he'd gone on up to level the site first, before finishing the road.

Don't over-clear your site. Open it up first a little at a time to see how far you want to go. Bulldoze only where it's necessary.

LAND AND SITE CHECKLIST

☐ Consider buying the largest piece of land you can afford.

☐ Buy land that you can't stand to let get away.

☐ Define your specific wants and needs and choose your property accordingly.

☐ Determine how you will finance and purchase the land.

☐ Camp on the piece of land you're interested in, or at least spend time there at different hours of the day.

☐ Choose a site that is not too steep.

☐ Run the driveway in first.

☐ Follow the contours of the hillside for your road.

And he'd pushed over all five beech trees. We did not build on that denuded site. The scar on that hillside can still be seen today.

In order to maintain a near level, follow contours of hillsides for your road; ditch and use culverts at low places. A good road should be functional and still have that series of unusual surprises, as natural as possible. A four-wheel-drive vehicle is almost necessary if you don't or can't level much, but they're expensive.

A pickup truck with snow tires and some weight in the back will usually get you in and out, and they're so handy. I once owned a very narrow, very old Jeep pickup with a minuscule engine that wound in and out of our trees with some agility and all sorts of cargo, as long as we didn't hurry. And not hurrying is something you must learn in this business of homesteading.

Site problems, from a money standpoint, blow up your budget. Consider stone ledges, marshy land, and steep slopes, which are all hazards. Every obstacle, however scenic, increases your cost in money or time, taking away from what you have to spend on the whole project. I do not agree with the modern builders who level everything first, or build in a flat field, but they do save money that way.

Site selection is not usually so complicated, but it gets more limiting if you must be near fields for farming, or near available electricity, water, a road, or a job. But in general, it must be a place you want to be, to look out over and to come home to. It should be a place you can disfigure as little as possible, leaving it as close to how it was before you came. And you must still love it.

Now, plan to spend a lot of time at your chosen site or sites before you build. Camp there. Wake up early enough to watch the mists lift from the hollows. See where the sun will rise and set. Can you stand that wind in winter? Will August bake you here? Get to know the moods of the place.

And finally, when you can feel that this spot is home, start visualizing your house on that site. And keep in mind that it can be only what the site will allow. A good site welcomes a house. A log house fits as if it grew here, of the materials that are naturally here. An artist friend of ours said that our log house reminded her of a mushroom. That's close, because a mushroom is never out of place in the woods. Nor is a stone or a tree. Work toward that, and see how close you can get.

CHAPTER FIVE

Design

YOUR HOUSE NEED NOT BE dark or dreary. It can have all the amenities, including air-conditioning and inspired plumbing. It can and should be for year-round living. No matter what the design, it will have a special feel and a mood like no other house.

Traditional designs age well — everything inside may appear old and mellow, or be a statement in sanded and oiled wood. Owners and decorators have transformed my restorations and new structures into imaginative living experiences. I like that approach best: the look of antiquity outside and whatever you like when you go through the door.

Hewn-log houses, weathered gray in rain and seasons of sun, cracked and aged in old axe marks and lichened stones, blend into the mountains and the fields and the hills as a fact of history. They follow a few basic designs that have endured as the most useful.

Harmonize with the Setting

Of course, your own house should be a personal statement, not necessarily an established building style. But architecture should fit, growing as naturally from the ground as possible, using natural materials in a harmonious, serendipitous complement to surroundings. If you choose variations on the basic designs — and you will, as surely as did the early settlers — let them be subtle, in keeping with the use of surrounding materials and craftsmanship. You will approach your finished house many times, and you must always be happy with what you see.

The classic log house was of hewn logs, V-notches or half-dovetailed notches, shake roof, stone foundation

Traditional log styles can adapt easily to personal requirements. An overriding concern is that the house should fit its surroundings.

Log house interiors need not be dark. These rooms were lightened with the addition of windows in the low knee wall.

This 54-foot house required an end-to-end log extension with interior log bracing partition. It was lengthened further with a frame addition on the end. The unevenly spaced dormers were strategically placed to allow light in critical areas inside, at the entrance, the kitchen, and the stairwell.

and chimney, raised wood or puncheon floor, pole rafters, small windows, maybe a loft and lean-to, and eventually a roofed porch. Most of the wood was worked by hand, although rough-sawn lumber, available generally for over a century and a half, was often used.

This house had a transverse ridge (at right angles to the front door), with the chimney usually at one gabled end and the door near the center of the sloped-roof front, which was generally longer than the gable dimension. (Public buildings — schools, churches — often had the front door or doors under the gable, possibly because they rarely had fireplaces. The peak also made a nice place for the belfry, up front and on display.) This was the basic frontier cabin, with lots of variations we'll talk about later.

Farther south and north, and west of heartland America, you'll find round-log work, different notching, different woods, reflecting different backgrounds and customs and availability of materials. And everywhere you'll see modern adaptations and slipshod imitations, with their inevitable picture windows and green plastic skylights. I shudder.

Don't jar the landscape. Look for natural lines, such as the slope of a distant hill or framing by trees. Pitch a roof a little more, or less, to blend with the terrain as well as to give you the upstairs room you need. Consider switching the chimney to the other end where the road winds, to catch it in your first view. Watch how the sun hits your site, and put more or fewer windows accordingly. Raise your foundation to give yourself a view into the next valley. Or position your house so your most used room will give you that vista you love. Your house should look as if you expected to find it there, just as it is.

Log House Size

The basic log house is traditionally 16 by 20 feet or larger. It's a rectangle, longer across the front and back to balance out an added lean-to or porch. But it's pretty small, as we view houses today. Full-length logs are necessary to bind it all together above and below windows and doors. And, like the pioneers, you'll find it difficult to locate, transport, and handle anything longer than 20 feet. A larger house also takes a lot longer to build and costs more.

When you build larger with logs, remember you must brace the long spans with partitions, or the whole thing shakes when you slam a door. Don't go for a non-log design, then try to adapt it to log construction. Learn from history.

A loft is very little more trouble and, if high enough, doubles your floor and storage space. Lean-tos, or end additions of kitchens and now bathrooms, can be built along with the basic cabin or added later, just as the settlers did. The size and shape of such an addition, like a porch, is a matter for your eye in relation to the rest of the house. It shouldn't dominate the house itself.

In the past, styles often changed with priorities. The pioneer built the one-room cabin with fireplace, then, when the family expanded, built another, separate one with its own chimney at the opposite end, and maybe joined the roofs to make the breezeway or dogtrot house. Sometimes both were built together if the family needed the space and had the time. One section was for cooking and sleeping; the other was for general living and sleeping. The upstairs, under the not-very-tight shake roof, was for sleeping too. Our forefathers had lots of kids, and sometimes maiden aunts and widowed grandmas, so sleeping space was a prime concern.

These dogtrots were sometimes two stories, with an upstairs porch running the length of the house, over the open area, and were large, substantial farmhouses. Curiously, the two front doors of the dogtrot have survived this style house in the mountains. Logically, each separate cabin needed a front door, and the dogtrot retained both, even with additional doors to the breezeway. But whereas the builders of the central-hall house (which may have evolved from it) favored an impressive single entry, the mountain house clung to the separate front doors. A notable variation is the 1870 Grigsby house, restored on the campus of Arkansas College, in Batesville. Its entry doors open on the breezeway.

Later builders abandoned log construction, closed in the breezeway, and even did away with it as a hall or room, but kept the two front doors. I have a file of pictures of Ozarks houses, some built as late as the 1960s, with dual front doors. Scholars hasten to inform me that this practice was continued in churches and schools to separate the sexes with a few feet of wood. J. Frazer Smith suggests that, when company stayed over, this was also the case with dwellings. In all, this double-pen log house was and is a picturesque style. If you need the space, consider the one- or two-story dogtrot, and count on a lot of labor and material.

The saddlebag house requires less roofing, but it's difficult to join the two log pens snugly. It means building both sections at the same time, or a touchy business of mortising in same-size logs if the second

Logs and stones go together well in almost any combination. The two-story log pen forms the core of our house, from which wings and porches grew. Our stone kitchen wing is of triple wall construction with timber-frame support.

The two-story dogtrot house provides the most space of any of the traditional designs. The breezeway was sometimes enclosed for even more space.

section is added. The partition as used in the Whitaker-Waggoner log house referred to in the first chapter is a nice solution, although this is not a saddlebag house.

For unusually large numbers of offspring, the settler often added a wing to his dogtrot or saddlebag, or even his "I" house, sometimes behind one half, with a side porch. I've just finished designing one of these for a family who wanted all the rooms of their dogtrot on one floor. Quite organic and fitting, and historically accurate.

That often turns out to be a great deal more important to me than to the people I have built for. And although I urge regional and historic integrity in log house architecture, it need not dismay you. You need live primitively in your house only if you choose. Such niceties as bathrooms have become necessities in our modern culture, and properly constructed are unobtrusive, efficient, and relatively nonpolluting. The same goes for kitchens. I prefer to locate these more or less modern rooms in lean-tos or wings, to make it evident that they are part of the added-on construction, not jarring into the historic appearance of the house.

And insulation is a useful, invisible, and comfort-producing material that you'll find very helpful and economical. A number of other modern conveniences, such as electricity, can be built in tastefully without offending most purists. If, in fact, your log house is to be your permanent home, it's better to build in wiring, plumbing, and insulation than to add it when you come to the realization that roughing it becomes a chore, year in and year out. Besides, building codes require many of these modern amenities in most parts of the United States now.

I do not pretend that these conveniences are entirely in keeping with the Spartan pioneer cabin, of which we have built several. People managed without plumbing for thousands of years, and if you so choose, I applaud your decision. I just don't envy you on wintry nights.

The basic cabin, electrified and plumbed or not, remains the starting point whichever way you choose to add on. Planning your future additions is an important part of decision making. As one example, I had

The dogtrot or open breezeway was often used as a dining area in fair weather. Later versions, such as this one in Arkansas, had the doors opening into the passage.

planned to rebuild the transplanted log house we had in Missouri as half a dogtrot, adding the other half as the material and time became available but keeping the breezeway open (read "historically accurate"). My wife, who at the time had never seen a dogtrot house, decided that going up, across and down, or out through the weather in order to reach point B from point A was a masochistic trip. So we located the cabin in another spot that allowed a lean-to but would have been impossible to build a dogtrot on. After two extensions of the lean-to, both of us were sorry we didn't have more room to grow. There's a moral: Build where you can add, in one way or another, without removing select trees or parts of a mountain.

Look at pictures to decide on a basic style. Get in your car, on your bicycle or horse, and go hunt up some specimens. Ask around — the natives are used to cabin-hunters by now. Take pictures; visit ruins. Harass the history or folklore professors at the nearby college. Find out what you're doing before you invest part of your life in this madness.

Avoid precut cabins (and preplanned houses of any kind) as consumer rip-offs. Because they use second-growth, smaller trees and generally unseasoned wood, these cabins tend to shrink, twist, crack, and leak. You pay for the nostalgia of the log cabin mystique without the craftsmanship. Borrow features and ideas, none of which is new, but let your site tell you what and how to build. That also goes for stone houses, glass, cathedrals, and motels. You're going to live with those trees and that hill a long time, so don't "magazine-house" them to death.

Consider Basic Family Needs

But, of course, you will have basic needs. Your family size will dictate a need for so many square feet, divided into rooms of whatever description you require. The business of good design always means intelligently and artistically integrating the requirements of the occupants with the allowances of the site in a style of their preference.

If you have a washer and dryer, for instance, you'll need a place to put them. A separate laundry shed would preserve the authenticity of your log house, but you'd clock a lot of miles over a period of time, going there and back. So put them in the lean-to.

And do something clever with the space under the roof slope; give it an upstairs. Slip in a mini-room for your youngest. Give the child a dormer window to let

The 1788 Captain Beadle house in Virginia was an adventure in design. We increased the living space with two additions — a frame lean-to and another restored log cabin.

The Wintergreen cabin re-created the "great room" concept by extending the main log pen with a smaller log addition for the dining area. This required additional interior log wall notching but results in a pleasing effect. The main pen houses the living room, an open kitchen, foyer, stairs, and a bathroom.

the world inside. A dogtrot house has lots of upstairs space, so maybe a studio or sewing room, away from traffic, is in order.

Design for the Logs First

There is a myth among modern Americans that bigger is better. And although it is true that you'll use every bit of space you build, I prefer an efficient utilization of smaller areas. Up high on the cabin wall in the lean-to is a world of room for shelves, cupboards, and cabinets (even a half-story sleeping space for someone small).

A great deal of the botchery visited on old log houses has been in the form of additions tacked on at random. By all means plan your house in such a way as to allow sensible additions that will not destroy rooflines or overall appearance. Give yourself enough space, and use it wisely.

I should state here that a common mistake, for some perverted reason, is to try to adapt contemporary house designs to log house building. A one-level ranch cobbled together of logs is still a one-level ranch, and is just as ugly. Only now it has stretched the function of the logs to ridiculous proportions, requiring all sorts of expensive techniques to keep the contraption from falling down.

Although it is true that the log house kit industry has catered to this post–World War II design craze of

wide multirooms on one floor, the results are visually awful. Build those things out of plywood and sheetrock and concrete blocks, and let the logs be used for loftier purposes. I cannot count the times I'm asked to repair these experiments in logs. That is not what I do.

Logs as material lend themselves to one-room-deep designs; there is potential for lots of light from outside-wall windows. We've talked about extending the basic rectangle lengthwise with more log sections or other tasteful authentic materials. Rear wings and ells and lean-tos are all historically authentic, and you may well want to choose from whichever of these fits your needs.

In restoring the 1788 Captain Beadle house in Virginia, we added living space to the house with two additions. The original 18-by-32-foot chestnut-log, two-story house was the traditional two-over-two design, which housed a parlor, staircase, and bedroom on the main floor and two bedrooms and a bathroom on the second floor. The kitchen had been in a decaying lean-to, and a dark passageway connected it to an equally decaying timber-frame structure that housed other bedrooms. We started by removing both of these additions. In their place, we built a new frame lean-to and moved an 18-by-24-foot, one-and-a-half -story West Virginia log cabin to the site to serve as a second addition. The cabin now houses the master bedroom and bath upstairs, and the lean-to contains the spacious kitchen and dining room with a view of the lake and backyard. Neither structure dominates the original tall two-story building, with its nine-foot ceilings, yet both complement the old house and the site. Hewn overhead beams and antique wood tie the sections of the house visually. The unusual design paid tribute to the evolution of the house and also resulted in a harmonious whole. Captain Beadle's historic house had been documented by the Daughters of the American Revolution (he's buried close by). The National Register of Historic Places required that we document each replacement and addition for its registration.

We solved the space problem in our Virginia house by adding a one-and-a-half-story stone section to one end of the two-story log original. Then we attached a two-story post-and-beam addition to the other end. So now we have a tall, skinny, long house with lots of sun gain to the south, small windows on the north, but, at only 18 feet deep, not too many square feet. A two-story covered porch shades the front in summer but lets the low winter sun in the big windows. We did have to build two stairways to avoid going through rooms on the way to somewhere else, but it all works.

And even though there's more post-and-beam and stone than logs, people still call ours a log house. That's sort of a key. When we design and build more than the basic cabin, we make it the focal point, flavoring additions with the influence of the logs.

So your log section might be a great room (living room with fireplace at one end and kitchen at the other), with sleeping room or rooms above. Additional rooms could be in an addition of stone, post-and-beam, or even stud-wall construction.

Basements were common in East Coast houses, because cool storage was a necessity before refrigeration. We still do a lot of basements on hillside sites, but I discourage the practice on level ground. Where a high foundation is necessary anyway and drainage is good, a basement makes sense. But it must be sealed well, insulated, accessed, and often plumbed. All this makes it an expensive space. The dampness that's a continuing aspect of basements threatens to decay wood, invites boring insects, and is usually pretty dismal. They tend to be dark as well. I prefer lofts and additions. A loft lets you see out at eyeball level with the birds and lets in light and sun.

Whatever your design, let your log house building be a refuge for you and your family and an exercise in reverence — for the priceless land, for the traditions of craftsmanship, and for your pioneer heritage, no matter how many generations gone. Not everyone is privileged to feel the weight of an axe helve in his hand and his feet on his own ground. Go quietly into the woods and work in harmony with the trees, resisting the impulse to change and brand the earth.

Design Checklist

- ☐ Make your house design a personal statement and a harmonious complement to its setting.
- ☐ Keep the historic integrity of the traditional log house.
- ☐ Plan your design based on your family's needs and the site on which you'll build.
- ☐ Choose historically authentic uses of space, details, and materials.

CHAPTER SIX

Acquiring Materials

THE QUESTION I AM ASKED most about log houses by serious seekers is, where do I get the logs? That's a valid concern, given the quantity of material the log house requires. Just the names — logs, beams, joists, rafters — sound like massive sections of wood, and they are. And then you'll need doors, windows, interior wall covering, door and window trim, and all the normal building materials like nails, studs, insulation, and so on.

Some important decisions will be based on the design and mood of your house. If you want your house to look as if time has stood still, then you will need to search for and acquire recycled, old materials. You can make sure all conveniences or modern structural materials are kept hidden. If you want open beamwork, whether for decorative or structural purposes, you must get beams to suit your dimensions and ideas.

You get the point. It will take effort to gather the materials you need. Not all materials can be "store-bought" at the local lumberyard or scrounged from the salvage yard or come from a house you are about to reclaim or recycle. Acquiring materials takes work. And those materials will fulfill your plans.

Recycling

Almost 90 percent of our material is recycled. We reuse hewn logs from other cabins, tobacco barns, springhouses, and corncribs, even use hewn beams as logs. We reuse old beams for interior and structural beamwork. Mellow, aged lumber trims windows and doors. We reuse old flooring (carefully relaid in order to keep the wear patterns in the right places). Or we buy remilled lumber specifically cut for flooring.

First of all, the old wood is usually better, being for the most part slow-growth heartwood, which lasts longer and gives boring insects indigestion. The heartwood is the best part, whether cedar, oak, pine, or poplar. Second-growth trees, by contrast, are mostly sapwood, which is a good lunch for bugs and the fungus that is rot.

Age alone of the material means that the green tree has dried naturally and decided what shape if any it will warp or twist into — finally stabilizing. The mellow color of old lumber or old logs cannot be duplicated from any bucket of stain. Old lumber just has a different look and feel from new lumber or a newly hewn tree.

Excellent-quality materials come from unlikely sources. This barn supplied enough good logs to build a substantial house. Other reusable materials, such as beams, flooring, mantels, stairs, stone, brick, and paneling, can come from dismantling older and even disreputable-looking houses.

Tobacco barns are a prime source of good-quality logs. Most were not chinked, and the logs are often in better condition as a result (improper chinking often held moisture against the logs and decayed them). Also, with few windows and doors, barns provide more long logs.

Logs

We find cabins up mountain roads and along major highways; there are falling barns at even the slickest modern farms. I have contacts in five states who locate materials for me, and can usually find the logs I need in a couple of weeks.

Old cabins become the source for houses moved and set up complete or for replacement logs on other projects. We have combined several cabins to create a larger house. Many times we find rot in important logs, such as sills, over doors and windows or top plates. We try to replace those logs with age-appropriate logs of the same wood. If we cannot, we hew out new logs in the same wood.

There are a few exceptions. For one project, we needed chestnut logs to replace a former cabin on an 18-by-28-foot stone basement in Virginia. It took five months to find a cabin to supply them, and it turned out to have some oak in it, too. Most people can't tell the difference, but this client could — it was our third restoration for her.

Another time, for the elaborate skiers' cabin on Wintergreen Mountain in the Blue Ridge Mountains, we advertised for two months to find the 30-foot chestnut logs we needed. Found those in a dogtrot barn 100 miles south.

Woods

Chestnut has become the wood of choice for hewn-log cabins. Oddly enough, it was considered a weed tree

This log house with many excellent logs was found for a Georgia client by Jeff Harris of Vintage Log and Lumber. Finding such venerable structures requires a great deal of effort. With the advent of the Internet, entrepreneurs like Harris have a new way to offer their services.

by the early settlers, because it sprouted everywhere and grew fast. But it didn't rot, so it was used for everything from split shakes to fence rails, and the tannic acid–laden bark was often stripped from the standing trees for leather manufacture.

When the chestnut blight hit in the late 1800s, chestnut trees died by the millions, and an age of fine timber passed. Only a few remain amid efforts to revive the species through research.

Heart pine, actually the slow-growth virgin wood of the longleaf yellow pine, was the first choice of early cabin builders in the mid-Atlantic and southern states. The Pennsylvanians, and many of the transplanted Europeans in the upper states, preferred the straight-grained white oak. White oak has proved to be one of the best woods for recycling. With the

White oak logs for reuse in our Missouri cabin. The stone foundation was filled as the base for a flagstone floor. The vehicle may vary but anything is fair game.

lighter-weight chestnut so very rare today, oak is a good choice, although harder and heavier.

Heart yellow poplar is also a good wood. Second-growth poplar is mostly sapwood, but the older trees, with the olive green heartwood, lasted as well as almost any wood. Called the "poor man's walnut," this was used for everything from wainscoting to shingles to furniture.

Heart white pine, used frequently in West Virginia cabins and northern timber frames, is a more stable wood than most, is easily worked, and lasts a long time. The trees are straight like poplar, and often yield big, wide logs.

These days I see and work with over 100 log houses a year, including consultations, evaluations, and searches for materials. I see some being restored of wood that is just not worth it — insect-eaten sapwood pine or poplar with powder-post beetles actively eating away in it. Building around these kinds of problems for the sake of "authenticity," or because the builder or architect doesn't know a rotten log from a good one, is a disservice and a terrible waste of money. Bad logs limit the life of the house, as well as build in frustrations and disappointments. A log house rebuilt with substandard logs only adds fuel to the negative stereotype of grandpa's drafty old cabin.

I've set some rigid rules for reuse of old logs that are a pretty good guide. If you see pencil-lead-size holes in the logs, avoid them. The beetle grubs will have eaten the wood under the surface. If there is termite damage anywhere, chances are it's spread throughout the cabin. Pass that one up. If rot is extensive, from too-close ground moisture or a bad roof or failed chinking, don't take a chance on it.

Pine and poplar rot from the outside in, and what you see is often the worst of it. I've said this before, but it's important. Oak and chestnut get water in through check cracks and rot from the inside. A sound shell may have mush inside. Thump logs with a sledgehammer and listen for the ring of sound wood. A dead sound means decay inside.

Often old wood gets very brittle. An informal test is to let a log fall off the truck; if it doesn't break, it's good. Of course by the time it's on the truck, it's often too late because you've already bought it.

Barns are often in better condition than cabins because they usually weren't chinked. Improper chinking held moisture against the wood and often rotted it. The wind blew through open barns and kept the wood dry. Also, hanging tobacco leaves in barns discouraged bugs that would eat the wood.

I worked with a couple in Iowa using hewn barn beams we dovetailed as cabin logs. They had scrounged the state for materials, collecting from four sites to have enough for the job.

Sometimes we reuse large beams as replacement logs. Here Mike Firkaly splits a timber-frame sill with wedges. Next, the split faces are hewn with a foot adze to create two hewn logs for use in the restoration.

Good logs are often hidden under clapboarding and plaster on the exterior and under drywall and plaster on the interior. Vertical strips were often nailed to log walls to align the clapboarding or drywall. In 40 years of taking down log houses, I have found many surprising materials used inside to smooth out rough log walls: old family albums (probably from the wrong side of the family), religious posters, old calendars, many layers of newspaper and wallpaper, mud and horsehair, as well as split-pine lath and plaster.

Log and Beam Length and Size

Remember that few logs must be full length for a log house. Cutouts for windows and doors and chimneys mean most of the logs can be short, and thus easier to find. Sills can be pieced. A one-and-a-half-story, 18-by-28-foot house may require only the spanners over the doors and windows, which also carry the overhead beams, and maybe two other pairs and the top plates, to be the full 28-foot length. At the ends, you can use a couple of 18-foot logs up to window height and maybe three or four above. At the chimney end, especially if there's to be a second fireplace upstairs, you can cut out maybe four of these.

We often combine old logs from several sources. We always try to stay with the same kinds of wood, even if the old builders didn't necessarily do that. I've found chestnut and oak, chestnut and poplar, oak and poplar, oak and pine, and poplar and pine combinations in old log houses. The fact is, few people can tell the difference once the wood is weathered.

When we use old logs as building materials, we often renotch if the old ends are bad. We also put in more windows, which means more light and the need for fewer long logs. A good rule for log length is to stay three feet from corners with windows for stability. Also, it's not a good idea to cut through the top logs

(end logs or plates) for windows, or to piece them. That leaves tall stacks of logs with little to hold them in line.

Beams for joists are often too small for building code–approved reuse. We use them for collar ties, and hew out or find heavier stuff for overhead beams such as 4×10 or 6×10 beamwork. When something more massive is desired and on view, we often hew log floor joists or reuse warehouse or factory beams. When we can, we leave the new material out for six months or so, turning it so it ages nicely. If you don't have to look at the joists, use 2×10 lumberyard floor joists, 16 inches o.c.

Costs

Prices vary a lot. At the time of this writing, the cost is anywhere from $10 to $20 a running foot for hewn logs, more for chestnut and for very long or very wide logs. Heart pine beams are usually sold by the board foot (1×12×12 inches) at $3 to $5 per board foot. Recycled heart pine t.i.g. flooring, 6 inches and wider, brings $3 to $5 a square foot. Remilled heart pine is from $4 to $15, depending on quality, presence of nail holes, percentage of heartwood, and whether it's flat or edge grain. Oak and maple are usually cheaper, being milled from new stock.

Salvage Yards

There are architectural salvage yards in most towns of any size, and you can find beams and boards there for joists, flooring, and paneling. In fact, a new industry has burgeoned to remill fine old timbers into flooring, trim, and beams. Any timbers have potential, with or without nail or mortise holes. We trimmed out the Wintergreen cabin in "naily pine," which might not have worked well as flooring but worked very well as window and door trim. Some of these enterprising businesses are importing timbers from as far away as China. This wood adds a beautiful and interesting dimension to design options.

Architectural details, stained-glass windows, iron-work, and other interesting items from yesteryear are also available from architectural salvage dealers. Doors are plentiful and often good buys. For years, we reused old plumbing fixtures, but found that they tend to be more trouble and expense than they're worth. You can buy new, trouble-free plumbing that closely duplicates the quaint stuff.

Windows aren't worth it either, because of the time and labor required to rebuild them and make them weatherproof. You might, however, use a stained-glass window as a fixed window or pattern your new ones after an old one you found. And wavy glass can be retrieved from a rotten sash and reused in your new ones.

Flooring

That mellow look of old flooring can come from several sources today. Take down an old house and reuse the lumber either in order or as random material. Today, there are excellent sources of remilled flooring — for example, old beam lumber resawn to flooring dimension, as mentioned above. The companies that remill old lumber for flooring offer various widths and characteristics. There is even a variety with old, blackened nail holes, which is popular for paneling as well as door and window trim.

Old flooring is in demand, but wear patterns mean some close matching to reuse it. You can take down an old house, number the pieces, and nail them back in order. Most of it 100 or more years old will be 5/4 inch thick, so it can be reused, then sanded down to show

This splendid example of pioneer building was moved in the early 1980s from its original site in the Shenandoah Valley of Virginia, where documents proved that George Washington had surveyed the original land grant. The 20th-century owners had other plans for the land and were not interested in restoring this house, so it was given new life over the Blue Ridge in the Virginia Piedmont.

the grain. Heart pine was most popular, but maple, oak, beech, and even cherry can be found.

Always get flooring that has been kiln-dried. Our house has three-inch-thick heart pine from an old cotton mill floor that had been stored in a barn for months. The boards, up to 16 inches wide, still shrank with winter heat. The same goes for paneling and door or cabinet wood. Unless it's really dry — 10 percent moisture or less — don't use it, or it'll shrink.

Cabinet shops and specialty milling companies have moisture meters to check content. This is best done by driving the sensors into the wood to measure the moisture inside. Wood can appear quite dry but be swollen with internal moisture. With the supply of good, recyclable wood in short supply, you will probably be left with the choice of new material throughout most of your cabin.

Cutting Your Own Logs

If you're lucky enough to find a site with lots of trees, give some thought to cutting your own, selectively. Although this may seem like wood butchery, careful thinning opens your site and will allow the remaining trees to spread their limbs, seeking more sunshine. Logging your land does alter it, but in a few short years your woods will stop looking like a disaster area.

Remember, though, that you'll need about 45 logs anyway, including possibly a heavy beam under the floor joists, plus overhead beams. Then there are rafters for the main house and porch and probably an addition. And then there are the floor joists themselves. And that's just for the basic single-pen house. You may also want to have your lumber cut from logs at the nearest sawmill, a practice I encourage. So unless your trees are very tall and very straight, you'll get maybe one or two hewn logs and some boards from each tree. Take a quick count and see if you really want to cut those trees and clear that much ground.

It is more convenient, but more expensive, to have a portable band saw and operator come into the woods to cut your logs. They eliminate your having to load, haul to, and then haul back your sawn logs and lumber from the large mill. Portable mills may cost twice as much per board foot as circle sawmills, but the convenience is often worth it. Most such mills cut up to 20-foot lengths. Some operators charge by the board foot, others by the hour. Since band saws have narrower blades, they waste less wood in each cut. Sawn cabin logs should always be center-cut to avoid warping.

Warning!

If you've never cut timber, you should assume that it's about as hard to do properly as using high explosives. It is not simple, and it is extremely dangerous.

Here are the basics: The tree should be notched on the side it is to fall toward. Use an axe, saw, or both to cut about four inches deep. Then begin the actual cut on the opposite side, a couple of inches above the notch. Keep cutting until there's only an inch or so of wood left, or until the tree starts to fall. Then get your saw out and run like hell. Have an escape route pre-planned. If the tree is thicker than your saw's reach, cut in first from the side so the bar will cut all the way through.

The following hazards are common: The tree falls backward on you or your saw; the butt end kicks out as the tree falls (particularly if you've cut too much); limbs broken off the falling trunk are whipped back at

FELLING A TREE

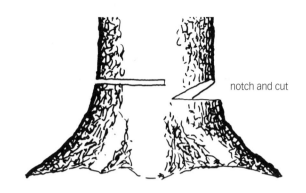

notch and cut

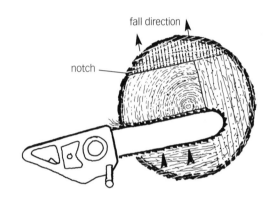

fall direction

notch

When a tree is too large for the chain-saw bar to reach through, make an initial cut on the far side of the trunk. Then proceed normally with the cutting operation.

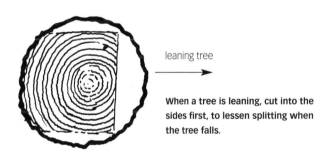

leaning tree

When a tree is leaning, cut into the sides first, to lessen splitting when the tree falls.

To change direction, leave more wood on one side, to pull the falling tree that way.

to change direction

you by the spring action of the surrounding trees ("widow makers"). Any of these can kill you.

If the tree leans much in the direction you plan to cut it, chances are it'll split before you've cut far enough through, so the butt will kick back as the split ruins your first log. Before you make the major cut, make three-to-four-inch-deep cuts on each side, and this will lessen the splitting. So will a last burst of frenzied cutting as the tree starts over, if you have the nerve.

Felling a nearly vertical tree where you want it, instead of where the wind or gravity wants it, is mostly a matter of aiming the notch — and the result of practice. Also, as you cut, leave more wood on the side you want the tree to fall toward. I have seen my father bring a tree a quarter-circle as it fell. My brother John delights in placing falling trunks between closely spaced neighboring trees.

A last word about felling trees: If your saw (cross-cut or chain) gets caught as the tree falls, spend a minimum of time trying to remove it. There will usually be a split second when it's free just before the trunk jumps at you, but don't bank your life on it. Saws can be replaced.

Beware of the attitude that "people do it all the time, so it can't be that difficult." That's just it — loggers do it all the time, and they get good at it. They learn to read a slight shift in the wind that can send the tree whipping backward before you realize it. They know which way broken limbs may fly, and what to do if a tree hangs.

I've felled trees for over 50 years, and I try never to be complacent in the woods. But for some reason, I recently failed to look up soon enough as a smallish poplar I'd cut started its perfect fall. It snapped off a dead limb from another tree 30 feet up, and a 10-foot length as thick as a baseball bat crashed onto my shoulder. I was on the ground with it and a running chain saw all over me fast — too fast. Two inches over and that limb would have hit my head. And my hard hat was in the truck.

I mentioned hung-up trees. Do not ever cut the tree that's holding up the other one. The combined weight will snap off the trunk before you've cut far, and the combined fall is fast and wild. Hook a long cable to the butt of the hung tree and use a winch or skidding tractor to free it.

If you cut your own nearby, skid them with a tractor, four-wheel-drive vehicle, winch, or patient mule. If you cut your own some distance away, have them hauled. More about this later.

A very independent-minded friend bought 40 wooded acres, fully intending to cut his own logs and build a log house. He bought a new chain saw and all the related gadgetry. Into the woods he trudged, axe in hand, wife in tow. He selected his first tree, aimed, fired — down came the tree, bringing three dogwoods down with it. His agitated wife said, "That's it. No more." Out of the woods they came. They drove to the sawmill and bought a truckload of 6×8 beams, then notched them to create a log house. He left the forest unmolested.

Alternatives

If you have no timber, your least laborious course is to visit that sawmill operator and offer to buy logs and lumber. He'll sell the lumber and usually the (sawed or unsawed) logs to you for a board-foot price (currently 60 cents to $1.00 per), or refer you to one of his log suppliers. These folks who sell to mills invariably own aging trucks, live out on faint roads, and haul a few logs to supplement whatever it is they live on. (I know; I grew up pursuing this very trade.) You may be eyed suspiciously at first, and you may have to wait a spell for your logs, but give this a try before you give up entirely.

Hauling Those Logs

Get your logs hauled to the site. Do *not* go out and purchase a rusting truck, justifying it as an eventual hay hauler, wood hauler, rock hauler, or cattle hauler. The damn thing will fall to pieces at all the tense moments, either mangling your body and/or delaying your building for many weeks. The same predictions apply to good old neighbors and friends who own trucks. Avoid them.

You see, I first learned to drive in a log truck in my tender years, and I have this battered observation: Logging takes the life out of a truck faster than any other business. My own tutorial machine was soon brakeless, starterless, and cableless. Even a pampered highway truck throws up its gears after a few bouts in the woods. An ex–garbage truck, that anonymously dependable remover of things foul, also turns mean under a shifting load of logs. The most docile flatbed rolling will seem to leap at stumps, mud holes, and precipices as soon as you declare your intention of burdening it with logs.

Have your logs hauled. You can get yourself killed suddenly, and of course permanently, fooling around with tons of loaded logs. Don't even help unload. Arrange to be gone when the logs are delivered, or find a pressing chore out of sight.

Still not convinced? Still eager to pit your mite of might against the timber (cutting, skidding, loading, hauling, unloading)? Then you've got a real ego problem.

I'll relate a simple logging trip from one of our log house jobs. You'll notice right off that I never take my own advice. But just keep your eye on the point, not the inconsistencies.

A Logging Tale

Three of us set out for a stand of near-perfect trees some 85 miles from the site (too far) and over 100 miles from our homes (entirely too far). We're minimizing the cost by taking a second vehicle, a pickup truck, that one of us will fill with free building stone at odd moments. Also, when our 19-year-old White log truck grinds to a stall, we can abandon it easier.

I drive the White, which after all those years of logging still looks like a garbage truck. My nephew Danny and a friend, Brad, who's working this summer to learn how to build a log house, go ahead in the pickup.

It is a July morning, before the heat rolls off the green flanks of the mountains. I coast down the twisting blacktop slopes, mentally anticipating the careening return trip, laden with logs inevitably too heavy and too long. The roar and vibration and fumes successfully numb my appreciation of the scene: hollows absorbing sunlight, the last mists blowing from the deep creek bottoms.

I have hiked this country and skimmed its rivers in canoe and kayak in less strenuous times. Now I am bent on traversing it, against what are usually long odds, with some of its beauty chained to my truck bed.

A load of cabin logs to be hewn and some sawed overhead beams and rafters for a new hewn-log house under construction.

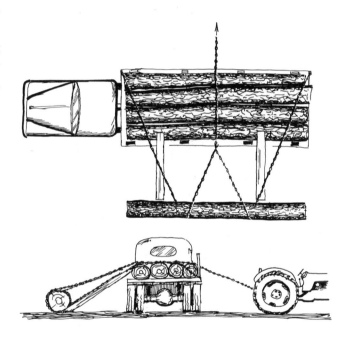

Cross-hauling with ropes or chains can load large logs. Smaller logs can be loaded by hand using skids or with a tractor that has a hydraulic bucket or boom.

So much for poetry, generally drowned in the howl of gears over three hours of steep inclines.

We arrive. I regain my land legs a step at a time. Danny is overhauling the magneto on the skidding tractor. Brad is measuring and marking trees we felled last time. They dry out and lose some of their weight in a few weeks. Clever of us.

The tractor refuses to start, but we pull it into life with the truck. I assault the fallen trunks with the borrowed chain saw (last time we felled these by hand with a crosscut). Brad starts to collect stone, mossy and aged, from the woods.

There's no hill or steep bank to load from, so we bunch the logs alongside the log road, then rig a cross-haul. That's one or two chains hooked to the truck bed, run under and around each log, and up across the truck to the tractor, which pulls the log up skids. A slow process. Some of these logs are 25 feet long, and some are 24 inches in diameter.

We have barely started loading when the tractor overheats, so we shut down for lunch. Linda has sent barbecued ribs (not a typical logger's lunch), which we devour. It is already afternoon, and heavy clouds are piling in the southwest. The tractor radiator is filled from a pond, and we cross-haul some more. It takes all three of us, two to shift each log on the way up as the big end gets ahead. Danny stops the tractor at a roared signal and holds while Brad and I strain with cant hooks to straighten the log.

Drops of rain start to spatter. We get the first eight logs on, less than half a load, as the shower stops. Now the skid trail is slick and the tractor spins its one bald tire. Danny brakes that wheel with jabbing motions, and the tractor does a series of uncontrolled lurches forward and sideways. It's uphill, and the nose lifts. That scares the hell out of me, because it can flip over on your body before you get your foot on the clutch. We clear a longer skid trail that's not as steep.

I go to drop more trees, tight business in this dense growth, when my brother John materializes. He'd thought the trip would be rained out, but later reasoned that we might be fool enough to go ahead anyway. John's an artist with timber, so I give him the saw and go spell Danny on the tractor.

Now the woods come alive. I use second gear and full throttle on the skid trails, with Danny clearing and hitching at the trees and Brad unhitching and bunching with the cant hook at the truck. John moves along the fallen trunks and limbs drop away; log sections roll apart. Everything is steamy and wet with sweat.

The next cross-haul takes us nearly to dark. We need more long logs to span over windows and doors of the house. John has them cut, and I head the tractor into the gloom as the other two finish the load of stone. Of course no one thought to bring a flashlight.

The last logs are skidded in darkness, John walking ahead, a dim shape guiding me around trees, stumps. The tractor has never run so well. The others wanted to stop an hour ago, but I suppose this is my macho trip. The long logs bend and grind and snap around behind my machine like alligators, but by damn they come out of the woods. Branches slap me; a long greenbrier gets around my leg and rasps its welts. John hurries ahead; I ride down a brush pile by mistake; the tractor sings.

We load by the pickup's headlights. It's a heavy load, dangerously long off the back, but we chain it down tight. Now we scrounge every drop of gasoline from reserve cans and even drain the tractor, which we will leave in the woods; no gas stations are open. It's a long, deserted stretch; we'll have five hours of grinding ahead. We check and add oil. We eat the last of lunch and some granola that John brought.

We plan to drive the truck in shifts — I'm first, then Danny. John follows me in his Volkswagen and the boys go ahead to wait and perhaps catch a nap at the Buffalo River bridge, less than a third of the way to our destination.

The truck is all dead weight on the two miles to the pavement, a half hour of slow churning. From there on, the long tail hangs out on curves too much and the front wheels want to paw the air. I stop and we tighten chain bindings. Gradually, I get to know the load, and the truck, ancient and massive, settles into its harness and pulls its rusted heart out.

We're on Highway 21 now, dropping down to the Buffalo River. This hill is 45 minutes of first gear, with the engine at a high singing moan, and a little braking on the steep places. It's like letting a weight down a cliff, this climbing out of the sky. I know the deep canyon below me; in daylight you catch your breath at its bluffs and distant waterfalls. Now it's all black emptiness, with maybe one isolated farmyard light way off.

At the last switchback I shift from first gear directly to top, rushing to full speed as I straighten out. I'm carrying nearly 20,000 pounds on a 16-foot bob truck. I must be crazy.

At the river we discover it's midnight, so we make some decisions. John heads for home and the cows he'll milk in a few hours. Danny, the shortest of us, crawls into the padded space up behind the seat of the White. Brad unrolls one of my sleeping bags on the broad hood of the pickup. And I stretch my aching bones on another, on the ground, with a poncho to help keep off the mosquitoes.

My trucks make ticking sounds, settling into sleep. I picture them both fleetingly, not as worn and rusted hulks, but as finely hammered steel, eager for work, sure under my hands and the hands of these good old boys working with me in this madness. Mosquitoes buzz, confused by the sticky repellent. Stars peep and the nearby river murmurs.

Sudden heavy raindrops pound the thin poncho into me and lightning silhouettes the trucks. Brad abandons the truck hood, doubling his length onto the pickup seat. I do not move, and soon am struck by a closeness to the elements, driven by rain that does not touch me, spread upon and flowing into the ground in fatigue. Come on, rain. It's just you and me. And these big trucks, and all these logs, and rocks. And pretty soon some more sleep, in spite of everything.

Morning sees us splashing river water in our faces, and then me and my truck grinding away up another mountain just as high. We stop at Kingston for breakfast, in a café in one of several old buildings put together, no doubt, with money from timber when this land was first logged. The high ceiling is embossed sheet metal, and everything smells wonderfully old.

Danny takes over, and the rest is July sun and curving hills till we reach the site. It's noon as we roll in, and it takes awhile for the landscape to stop moving.

We unhook the chains cautiously, because this simple operation can kill you, if those tons of logs storm loose before you're ready. We pry and roll the last of them off, and the truck straightens its back. I pat it on the nose as the boys roll two specimens onto cross beams in a sea of chips.

Then we reach for our broadaxes.

Well, that's an account of an actual logging trip. I could also tell you about the time we had a timber deadline to meet in January, and worked all night in a sleet storm, skidding and loading, to haul the next day on ice — but you get the point. Logging is *always* a lot harder and takes longer than you expect. And it's expensive. You see, the timber had to be bought first, at about 10 cents a board foot, and we paid ourselves (too modest) a wage. Add to that, five more trips for logs, both for the house walls and for lumber. And two big truck tires at over $100 apiece. And a tractor overhaul, and a clutch job for the pickup truck. And all those trips that took two days (or three, even) instead of one (as you had planned).

So buy your logs and have them hauled, unless masochism really is your thing.

Store-Bought Materials

We tend to use new material where it won't show, as in rafters to be boxed in, and floor joists and added-room stud walls. Beams, logs, paneling, stairs, and flooring are on display and we thus save the best old wood for them.

The logs are only your main cabin walls, so unless you have your lumber sawed from them, you still need many feet of boards. You can use rough-sawn boards and studding of yellow pine or poplar for just about everything in the house that isn't log or stone. If you'd rather, go to a lumber company and pay its prices for dimension stock. You'll find it more expensive but more uniform in size, making it quicker and easier to install. Rough-sawn lumber often varies a lot in thickness and width.

You can, of course, use poles or hewn beams for things like rafters, but it's a bit difficult to build a stud wall for a lean-to or upstairs room without sawn dimension wood.

Roofing

I use either standing-seam metal or split shakes on the roof. It takes so many shakes, you may want to buy these too. I just don't find enough good wood (white oak, cypress, red cedar) to split them realistically. A happy solution is the rare shingle mill that can saw cypress or cedar, knotty or not, at about what the lumber companies charge for theirs. We'll talk more about shingles in the chapter on roofs.

Don't bother recycling tin or shakes. Slate recycles nicely, and can be had in areas where it was common. Slate shingles are laid exactly like shakes are. If modern building codes are followed, roof framing will be strong enough for slate.

To build the chimney on our house in Missouri, we gathered stone from our property.

ACQUIRING MATERIALS CHECKLIST

☐ Recycle any materials in good enough condition for reuse.

☐ Look for logs in good condition; don't build unnecessary problems into the process.

☐ Check salvage yards for beams, boards, doors, ironwork, etc.

☐ Check wood for moisture content to prevent excessive shrinkage.

☐ Cut your own logs only if you're fully prepared for the danger and hard work.

☐ Have your logs hauled and unloaded by a contractor with the right equipment.

☐ Use new materials where they won't be obvious.

☐ Inspect all materials before purchasing or hauling away.

Stone

Stone may be free from a farmer who wants it out of his pasture, or $75 a ton from that same farmer, or more through a stone quarry. It can cost that much even if it's free to begin with. You've got to consider equipment rented, borrowed, fueled, and with breakdowns, other mishaps, and time figured in. Of course try to use any stone you find on your land. Stone recycles well, from old chimneys, basements, foundations, and walls.

Greenstone and granite are heavy and not easy to shape when necessary. Porous, crumbly sandstone is weak. The best dense sandstone is somewhat heavy, but good to work with — more so if you have ledges that break off in even, stratified pieces. I like that off the top of the ground, with lichen and a patina of age on it. Limestone is often available, but has a commercial building look to it. It works well, though. That's a matter of taste.

Obtaining Other Materials

Whatever you buy — be it stone, logs, or lumber — ask around for competitive prices, remembering that hauling from a distance eats up savings. Inspect logs, lumber, and all other materials. Folks have a way of loading you up with castoff stuff when they think they can get away with it.

Windows, nails, cement — all that material is about the same for a log house as for any other, and we talk about most of it in the following chapters. It's easier to find a large building supply house that can provide you with everything from sand for mortar to felt for window stripping, so you don't race around all over searching for such goodies.

No matter how meticulously you plan, you'll never remember or budget for everything. And in the year or so you've given yourself, many of your first ideas will alter as you come to realize the size of the project. You can count on extras everywhere; if you save a bit here, something you hadn't counted on will take it, and more, there. But in the end, you'll have the house you really want.

CHAPTER SEVEN

Foundations and Basements

LAYING WOOD AND STONE on the ground to start a log house was one of the things our pioneer forebears just weren't very good at. Invariably, sill logs were put too close to the soil, or right on it, which explains why so few pine cabins have survived, even with restoration. Termites love pine. And pine, oak, or even the long-lasting woods such as cedar, chestnut, and walnut eventually rot, and it happens soonest close to the damp earth.

Early Foundations

Many of the settlers' cabins were on hillsides, so only the uphill sills were on or near the earth, and maybe the ends of the first side logs. Those that were high off the ground lasted the longest, so take a lesson from that. It usually means a high floor, but it is worth it after a few soggy seasons. And building codes require 18 inches or more crawl space for plumbing, wiring, and insulation.

Some pioneers did build high to begin with, but the weight of the house, concentrated on a small area of foundation stones at each corner, eventually drove the stones into the ground, lowering the whole house. Certainly much of the sag and tilt of old log houses is due to this settling rather than to poor workmanship in the house itself.

Too few early builders put up a wide, heavy foundation. Most stacked flat stones singly, achieving a teetery column with maybe rock chips wedged into the cracks. Some cabins had pieces of metal from broken plow points wedged between the stones to give the piers stability.

Even stones laid on top of the ground, in a row beneath the logs with no footing in the ground, will give good support if there are enough of them, because, of course, each is carrying less weight and the total load is spread over more area.

Footing and Foundation

A house footing is the "foundation for the foundation." For a continuous foundation, the old builder's rule says the footing ditch should be twice the width of the wall thickness and at a depth below frost line. Even slender columns of stone at the house corners would be enough to resist settling, if laid on wide concrete

This drystone foundation was built for the restoration of the Sam Black tavern. The unmortared granite duplicated the original.

footings. So if you plan only a support column at each corner, you need more footing to be safe.

Fortunately, unless they built on creek bottom land, most of the mountain settlers found, only inches down, very rocky, firm subsoil that really resisted settling of any kind. This soil allows for a minimal footing, and will let you get past the slow, nonvisible stage of building sooner and on to the good stuff with the logs.

A continuous foundation for a log house just isn't necessary (although most building codes require it). The logs act as massive beams to distribute the weight

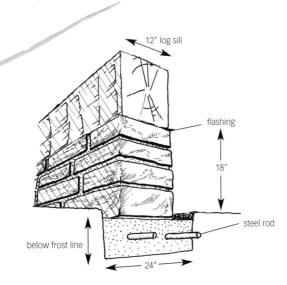

This cross section shows the reinforced concrete footing, stone foundation, and the sill log laid on a metal flashing termite shield. Floor joists were traditionally mortised into this sill.

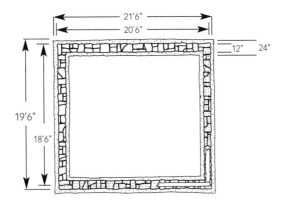

When laying out footings, allow extra space so that the foundation wall is centered on them. Foundation corners with midwall support are structurally sufficient for a log house, but building codes may require a continuous foundation like this one.

of the house evenly, so support can be at wide intervals. The benefits of a continuous foundation do include a way to keep winter blasts from under the floor. And it's hard for mice and other creatures to steal your insulation.

Muskrat Murphy's cabin foundation near Hilda, Missouri, is continuous, laid on a steel-reinforced, foot-wide concrete footing below the frost line, with ventilation on all four sides. Most building codes do not require the reinforcing rod, and only specify vents front and back, within three feet of each corner. The screened vents are closed in cold weather.

But there's good storage under there, and the floor stays warmer in winter.

We build workshops and storage buildings simply on 3-foot corner angles of 12-inch-wide stones, laid flat on just enough concrete below frost line to give even support. We'll usually build another set of piers midway from the front and back walls, too, if the house is over 16 feet long. There's sometimes an additional pier supporting a sleeper under the floor joists, halfway down the sides. This also gives support to the walls midway, which is a good idea if your cabin is more than about 18 feet deep.

In yet another variation, we will build on stone corner piers set on reinforced footings for owners who want to fill in a continuous foundation later. We dig and pour a footing all the way for their later use.

Filling in between corners is a simple matter of laying stone up from a below-frost trench till it reaches the bottom of the sill logs. The fill-in carries little weight, but seals out weather and creatures. It can be done a little at a time as you collect stone, and it sometimes happens that the house itself is almost complete before the foundation. We did the Bennett Alford house in Nelson County, Virginia, this way, which was later filled in with stone from their woods.

The log house my family and I live in has a continuous foundation because we have a flagstone floor set on fill, with the logs being 18 inches up from floor level. There is a reinforced footing because we hit solid rock in places, and natural, uneven settling of the house would crack the footings and foundation.

For your own purposes, you should build a longer, wider foundation for softer ground, laid on a reinforced footing. Of course you should use stone, if possible, and without the footing, settling will crack your masonry. With the logs above to distribute and bear the weight evenly, and reinforced concrete below, all the stones will have to go down together if they sink at all, and they won't sink much.

Settling by degrees is a fact of old buildings, unless they are on solid rock. That's all right today, too, if all of your foundation is on the same rock layer. Just one corner on soil will cause the house to tilt, in time. So that's why we reinforce the footing. In addition, if you've had to dig out a stump, you should fill the hole with broken stones and reinforce the footing that goes across the hole.

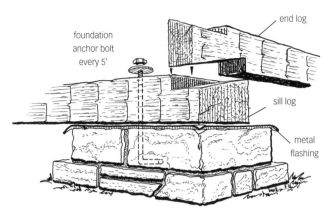

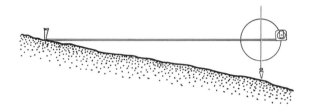

Sometimes we build only a stone corner pier with a footing so that the continuous foundation can be filled in later.

Footing Ditch

Lay out your foundation ditches with stakes and string, allowing for the extra width of the footing. If the outside dimensions of your house are 18 by 20 feet, you'll set those 6-inch-wide logs on a 12-inch foundation, set in turn on a 24-inch footing to below frost line. So you'll have a total of 19 feet 6 inches by 21 feet 6 inches, with the ditches extending inward 2 feet. You'll want the outside of the foundation flush with the outside of the log wall.

In measuring this out on a hillside, use a plumb bob or simply a weight on a string to get the distances measured accurately horizontally, not at a slope. Once you've laid out the square or rectangle so that the sides are the right dimensions, you'll need to check for square corners. You can sight with a carpenter's square to get close. Or measure a given distance from each corner to get a right triangle, square each leg, add, and extract the square root to see if the hypotenuse is correct. If not, shift two opposite walls. The formula is: 6 feet down one wall, 8 feet down the other, and the hypotenuse should be 10 feet.

I lay out the sides by simply measuring from one corner to the opposite one, then measuring between the two remaining corners. If the distance is the same, the corners are square. If not, I shift walls until the diagonal distances are the same. Be sure the outside measurements are right first.

Standard procedure then is to mark the footing ditch lines on the ground using mason's lime. It's so white, a sprinkled line of it will show up clearly. Even

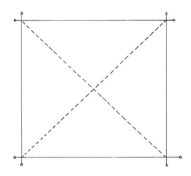

When laying out a foundation on a hillside, use a plumb bob to get accurate measurements. Irregularities in measurements here can create problems later.

After laying out the rough square or rectangle for the footings, check corner to corner for square. If the diagonal measurements are not equal, shift two sides until the diagonals match.

a backhoe's shifting around usually won't wipe out the lines before the ditch is completed.

It's really masochism to pick-and-shovel your ditches, although I've done it many times. Backhoes are around $55 an hour at this writing in central Virginia. A good operator will have your 18-by-24-foot or so ditches dug in around four hours — more if you

Where a hill slope allows, a walk-out basement is a practical addition. This English basement (which normally houses a kitchen) was a restoration, requiring that the stone wall be stabilized and sealed from the outside to preserve the drystone look on the inside. The warming oven restored in place indicates that the original was used as a summer kitchen.

want him to haul off or spread out the dirt. I like to keep the topsoil in one pile, and just enough subsoil for backfill later. Don't let the backhoe operator spread clay over your lawn area if you want to try to get grass to grow in it.

If you're some distance away, most operators will charge you travel time, especially for a small job. It's a good idea to have the same man install your septic system at the same time, and maybe your driveway, to save on his machine travel time.

Renting a backhoe for your own use is costly (around $150 a day for a very small one), and you won't get much work done until you learn how to operate it. You can get yourself or someone else hurt with all those hydraulics, too. This is definately a good time to hire a professional.

Your continuous footing ditch will require inspection. The inspector will check ditch depth width, bulkhead boards (which step the footing downhill), and grade stake height. Grade stakes are usually short

lengths of rebar driven into the bottom of the trench every four feet or so to indicate the depth of concrete to be poured for the footing.

Generally, a one- or one-and-a-half-story cabin will require eight inches of concrete footing. Some inspectors want rebar in it, some don't. I drive the grade stakes down and check height with the transit or a four-foot level. The idea, of course, is to have the footing level, with even steps downhill. A sloped footing invites the foundation, and cabin, to crawl downhill.

No matter how careful you are, you will find that by the time you've brought your foundation up three feet or four feet to level, you've let at least one corner get out of line. The one-foot width of the stone foundation wall gives you three inches each way for error in locating the log wall, and you may need it. Check your first course of logs for square, too, and slide them around on the stone till they're right. Building codes require that the sill log be bolted down with foundation bolts set in the masonry, so get everything square as you go.

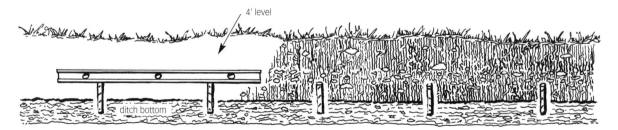

This is the foundation ditch. The footer will be of poured concrete, to the depth of the grade stakes.

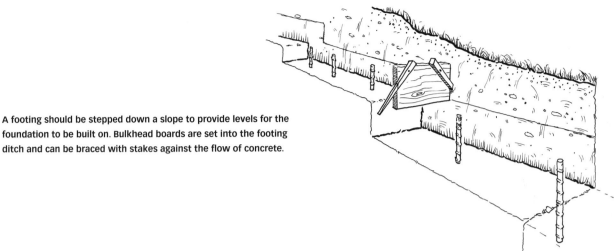

A footing should be stepped down a slope to provide levels for the foundation to be built on. Bulkhead boards are set into the footing ditch and can be braced with stakes against the flow of concrete.

Basements

If your site is on a hillside, you may want to look into the possibility of a basement. I don't recommend them on level ground because drainage is always a problem, and sealing is trickier when groundwater stays against the walls.

For a basement, lay out and dig at least two feet of extra space each way. A track loader is the machine of choice for digging out all those cubic yards of dirt. You can have the same operator build your driveway while he's at it.

The footing ditches can then be dug, preferably with a backhoe (most excavators have both machines). Just lay out and dig as if you were on flat ground. If you've hit solid rock, which often happens in the mountains, you can form up the footings and pour them on top of the ground, then fill in with crushed stone up to slab floor level. You'll backfill, so all this will be below frost line.

Basements today are poured concrete or concrete block. Building codes require 12-inch block below grade, which can be stepped back to 6 inches or 8 inches aboveground, and faced with stone or brick. If concrete is poured, have the forms stepped at grade for the ledge to set facing veneer masonry on. Or if perhaps you plan to coat block or concrete with plaster or stucco, leave it full thickness. Foundation bolts should be set every five feet for the sills, to anchor them in place.

Building concrete forms is expensive. I hire concrete men who use reusable fit-together forms. I prefer the poured concrete; it's stronger and sometimes cheaper. But with block, corrugated masonry ties can be put in to bind the stone or brick veneer. With poured concrete, we use a masonry nail gun to shoot the ties into it.

Basement walls must be sealed with one or two coats of masonry coating, such as Thoroseal, which, with at least one coating of tar, will do the job. (Use three coats total.) There are professionals who'll do this for you, even including insulation, for a surprisingly low price. As a contractor, I'm continually comparing what it costs to have my crew do such jobs, and it's often cheaper to hire someone else to do them.

The Floor and Drainage

Now, you must lay perforated drainpipe around the poured or block basement walls, leading downhill to daylight or to a French drain (a hole in the ground filled with broken stone). Cover the pipe with at least a foot of #6 to #8 crushed stone, and cover this with filter fabric, which lets water through but not silt. Now it's time to backfill. After a few weeks, the fill will sink, and you'll have to top it up again with the topsoil you saved from excavation.

Basements usually have the plumbing pipes in a concrete slab floor, so these must be laid out and installed before the slab is poured. Sometimes the slab is poured later, after the house is up on the basement walls, but usually all the concrete is done at once.

Before the slab is poured, a base of at least four inches of crushed stone goes down, and a vapor barrier of six-mil sheet plastic. Some folks put Styrofoam insulation under, too. Then 6×6 welded reinforcing wire goes down before the pour, overlapped so there's no open space between runs.

LEFT This is a precast basement set on compacted gravel. The panels are bolted together and sealed with element-proof adhesive. For this basement, we required a ledge for the stone veneer that covered the aboveground part of the foundation, as well as an allowance for the stone chimney for the woodstove. When this restored log house was finished, it still looked like a log house on a stone foundation.

ABOVE A poured-concrete basement is set on a traditional concrete footing. The floor system here was engineered to minimize flexing in the 26-foot span.

Four inches of concrete is generally enough, but if there's to be heavy machinery or a garage, go to six inches. Make sure the reinforcing wire is pulled up to the center of the slab as you pour, because it does no good down at the bottom.

Concrete slab work is best done by professionals, who can leave you a smooth, level surface in a short time. Unless you've had experience, a slab will give you nightmares. You can get everything level using a screed board, but smoothing must be done with trowel work.

Basement Walls

If your hillside is steep enough, yours becomes a walk-out basement. If not, you should cut the downhill slope longer to get down to below slab level. Steps down into a hole to get to the basement invite drainage and moisture problems. A walk-out allows windows for light and the view, and generally makes the basement more livable or usable for shop or storage.

For living space, basement walls are usually finished with drywall over Styrofoam insulation between furring strips. The strips are nailed to the walls with masonry nails (which can be driven by hand or with an explosive-charge nail gun) and two inches of the foamboard insulation glued in place between them. Board paneling can be used instead of drywall, but it

is more expensive. Few of the people we build or restore for want drywall, so we shiplap wide boards, say 1×10s, and paint them. To reduce cupping, it's always a good idea to prime or seal the back surfaces before nailing up.

Restoring Basements

Traditional stone basement walls have a special beauty, but they are damp, leaky, and often give way under hillside soil pressure. We've restored these by digging down outside, pouring concrete against the stone, sealing, drain-piping, and graveling to leave the inside as it was. This is expensive, but it preserves the original good stonework and is pleasing, too. The Page Meadows house near Charlottesville, Virginia, required this technique, because the old basement was dry laid. We formed up for the pour, but did it in successive phases so as not to exert too much pressure on the stone walls.

Restoring basements can be tricky. We've lowered basement floors in existing houses, breaking up the slab to dig it deeper. On the David Smith restoration in Virginia, we had to take out enough wall to get a loader through. We also put a footing under the stone foundation there, with the house atop it. This involved digging under the foundation, a two-foot-or-so-at-a-time operation. When each section was poured and set up, we went on to the next, leaving rebar extending to tie the sections together.

A contractor I know once dug out too much for this operation and the basement wall fell in, leaving a hole you could drive a truck through. Brick damping had so weakened the basement walls that, as a result of this mishap, the owner decided to demolish the entire 1840s house. It wasn't a log house, but a huge, beautiful Italianate-style brick plantation house, which cracked and settled in a way logs would not have.

Preparing for the Log Setup

With either a basement or a continuous foundation, you can actually do away with sill logs, which generally support the floor joists. Historically, the joists were notched into the sills. Today, many builders use galvanized metal joist hangers. You may leave a wide foundation, with the first log at the outside and the ends of the joists on the inside, resting on the masonry itself. The first log is below floor level, as is the sill, but it doesn't have to support the floor.

Bring your foundation up a foot or more off the ground before you lay the first log. Eighteen inches is better, because termites don't like to travel far from ground moisture. And you want that much crawl space. After the logs are in place, fill in any gaps to give more support, using mortar and stone where needed. And use metal flashing between the foundation and the first logs.

I am often required to set heavy bolts into the foundation masonry to anchor the sill logs. This is commonly done in conventional stud-wall construction and in post-and-beam construction. It would be of no value in a log house, however, unless some way were devised to fasten all the logs together, from the foundation on up. The weight of the logs themselves holds the house together, and I have never seen a dovetail-notch cabin blown apart by wind. (The roof might depart in a really heavy gale, but we talk about how to prevent that in the chapter on roofs.) But building codes often require anchor bolts. And they do help keep the sill logs in place laterally.

Set first logs and joist ends on metal flashing or on pressure-treated wood (building codes allow either) to

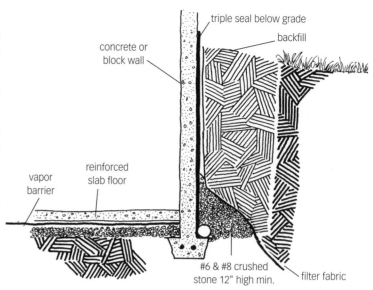

Perforated drainpipe must be laid around a basement wall below floor level to carry away groundwater. Gravel and filter fabric are necessary to keep backfill soil from clogging the pipe.

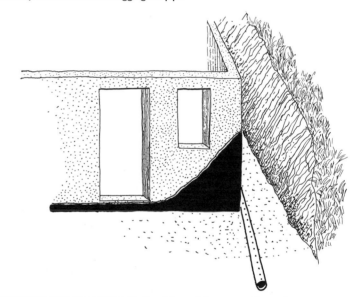

Basements must be sealed against moisture all the way up the walls, which will be backfilled. Drainpipe, in conjunction with the sealant, will keep water from soaking or seeping into the basement walls.

foil termites and stop moisture damage. Moisture wicks up from the ground, through the foundation, and the flashing stops it. The treated wood doesn't stop moisture penetration, but the poison of the treatment kills both the rot fungus and hungry termites.

Aluminum or galvanized flashing will work, but we use copper. It ages into invisibility and, being thicker, lasts longer. But it costs more too. (Depending on the market rate per pound, it costs from $200 to $300 for a 2-by-50-foot, 100-pound roll.)

This dry-stacked stone basement wall on a major restoration leaked with the April rains. The new owners wanted to have a functional basement and retain the dry-stacked stone wall look on the inside. To waterproof the basement, we dug below floor level to expose the outer wall and installed perforated drainpipe, then sealed and insulated with a three-inch layer of urethane foam, installed filter fabric and gravel, and replaced the fill dirt.

Here, the owners tried to retrofit a stone foundation on the outside of the log wall. This will not work without correct installation. The rain splashes back, damaging the logs, and the stone veneer pulls away from the wall, allowing water to get behind the stone.

The Concrete Pour

Your pier-footing concrete is best mixed by hand or with a small mixer, because you'll use so little of it. If you pour a continuous footing, plus a pad for the fireplace, you should buy ready-mix from a supplier. Its trucks are big and heavy, however, and you'll need a pretty good road in to accommodate them. I'm partial to narrow, winding, even steep approaches to sites (whatever the contour dictates), and few concrete trucks can get to such places easily. We've had to pull or winch them up steep places.

Economically, you won't save much by mixing your own. For the footing and pad, it's easy to spend more for mixer rental, cement, hauling, sand, gravel, and water than ready-mix costs. Most suppliers have a minimum load, though, so make sure you'll use it all. And there's usually a travel charge for long distances.

If you buy ready-mix, ask for 3,000 p.s.i. mix. If you mix your own, use a 1:2:3 proportion of Portland cement, sand, and gravel. Try to get gravel one inch or less in size. At no time should the thickness of the biggest stone in the mix be over half the thickness of the wall or slab to be poured.

Don't mix too thin. The water will evaporate, leaving air bubbles that weaken concrete. Mix wet enough to get all those dry pockets of sand or cement that build up in the mixer. And try to mix and pour a complete job at a time. Cold joints are weaker, so pour the complete footing or slab at one time.

A footing on a hillside should be poured with steps in it, to give as near a horizontal surface as possible to lay stone on. Pieces of board (bulkheads) wedged on edge into the trench at intervals do nicely. A lot of low steps (eight inches) work better than a few big ones, so use pieces of eight-inch board. You may want to use solid or filled concrete blocks to fill up to ground level, to save stone. In this case, the eight-inch steps work out right.

Reinforcement

I prefer half-inch reinforcing bars, and use two of them in a footing eight inches deep and two feet wide. Pour half your depth, then lay the bars, side by side, four inches apart and four inches from the edges, overlapping the ends. Take time to bend the bars where you need it, to keep them near the vertical center of the footing. Then put the steps in, if any, and pour the other half on top.

For ready-mix, you should have the reinforcing rod and bulkheads in place. I use short lengths of bar driven into the bottom of the trench, along with the grade stakes, and fasten the long bars to them with tie wires to keep them up off the ground. Let concrete cure at least two days. In freezing weather, insulate over the pour with straw or sawdust to keep in the heat.

Other Foundations

There are a few other options for building foundations. A low-cost foundation can be constructed from concrete block on a footing. You can leave it as is, if you don't mind looking at it, or you can parge the block with mortar to hide it. A poured-concrete foundation is strong but ugly, and requires form work. Both can be stone- or brick-veneered, set on the footing. However, building codes do not allow for the building weight on the veneer.

A more expensive, but innovative, foundation is the precast dense concrete foundation that is custom designed for each house. The entire basement for one log house we built was installed in three hours. Installed with a crane, the 10-foot panels come with Styrofoam insulation and studs predrilled for electric wiring. The panels can include basement steps, foundation for a chimney, ledges for stone or brick veneer, ledges for the floor-joist system, windows, and doors.

Covering exposed concrete foundations with stone veneer requires the use of masonry mortar (which is different from footing mixture). I use a mix of one part masonry cement to three parts sand, and I use one part lime to two parts Portland to make the masonry cement — that's one part lime, two parts Portland, and nine parts sand. There are lots of other formulas, but this has worked nicely for me over several years. And it's cheaper than buying masonry cement or premix.

Mortar should be thoroughly mixed in a box or wheelbarrow with a hoe, using enough water for a slightly stiff mix. Most masons use it as dry as possible to avoid smearing and running, but it bonds to the stone better if it's wet (just dry enough not to run).

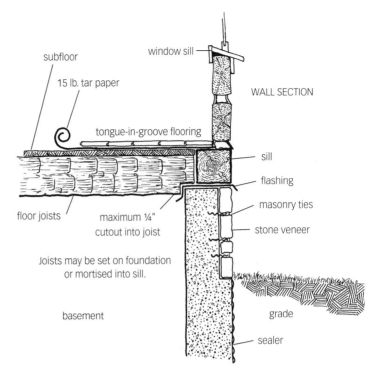

The floor joists can be set directly on the foundation ledge inside the first logs, eliminating the need for attaching to the sill. Both the sill and the joist end must be set on either metal flashing or pressure-treated wood to act as a termite shield.

I don't mix masonry mortar in a cement mixer unless I have plenty of help and plan to use many batches. You'll find that if you work alone or with one helper, it takes quite a while to use up a couple of cubic feet of mortar, and you may mix but one or two batches a day. Also, a lot of it gets caught in the tines of the mixer and wasted when you wash it out.

Both concrete and mortar must be kept moist for several days to complete the chemical reaction that produces strength. Just letting it dry out doesn't work. Wet burlap sacks are good to cover fresh work, or sheet plastic. Spray with water, but not until the second day or you'll wash it away.

And just don't do mortar or concrete work in freezing weather. Portland cement generates a little heat the first day as the chemical action goes on, but it's only good to about 30°F. It's best to use insulation and sheet plastic to hold heat and ground temperature next to the work the first two cold nights.

Below 28°F or so, just don't lay stone or build masonry aboveground. The chill will permeate even insulation in time, which will freeze your work. Footings down in the ground, with some straw over them for insulation, can be poured safely down to about 20°F. I don't use antifreeze solutions in mortar or concrete.

Large stones require no more time to lay than small ones. This mortared stone foundation was for a log blacksmith shop restored as a cabin on Greene Mountain in Virginia. The space for a foundation vent and the metal flashing between masonry and sill are building code requirements.

Drystone foundations allow air to circulate under the house, which keeps down moisture but makes the floors colder. Drystone requires no footing or mortar but must be carefully laid to withstand flexing from freezing and thawing earth.

Stone Foundation

Now, about the stones in the foundation. First of all, haul in more than you could possibly need. You'll still run short. If you can, get stone that's flat top and bottom. The face is a matter for your own taste, but structure is important here, so lay as much as possible horizontally. I've benefited from living in the Arkansas and Missouri mountains with the best building stone everywhere for the picking, and I've lived in Mississippi and in central Virginia where rock gathering is a long-distance adventure. Use a little mortar on the footing concrete, maybe one inch thick, wetting it first, just as you will for each course or layer of stone on previous drywork as you go up.

Lay the stone any way you choose, but cover the joints between stones with the next layer, brick fashion. You invite a cracked foundation if mortar joints align vertically. I like to lay stone flat, or ledge, with the edges showing. I vary this with a square, rounded, or angled stone occasionally. A wall laid with the stones standing on edge is weaker, unless very thick.

Make sure each stone will stay where you want it before you mortar it. That way it will be there from now on, even if the mortar erodes away, as did the mud, clay, and lime mortar of earlier foundations and chimneys. Lay about one inch of mortar, back from the face an inch, then set the stone, rocking to ensure

bonding. Then trim any excess mortar. Rake the joint at least half an inch deep so the faces of the stones project. Try not to smear the mortar. You can clean up the next morning with a wire brush.

A drystone foundation is more nearly authentic pioneer, but is also a haven for snakes, and is often not allowed by building codes. One advantage is that it's less work, because no mortar is used and no footing is necessary. Water can get under the stones and freeze anyway, so there's no point to a footing below frost line. Most of the water drains out from between the stones naturally. This type of masonry requires good, flat-surfaced stones, however. And that means a lot of searching for the right rock, wedging between, or a lot of stone cutting and dressing. A foundation obviously shouldn't totter, or folks in your house will tend to become uneasy. So wedge stonework with bits of rock or metal. These shims should always be flat, or flexing with temperature changes will drive them out.

In the process of the ground's freezing, swelling, settling, and washing, you'll notice some changes in your drystone foundation; but if your stones are good and wide, and the piers cover a lot of ground area, you can live with it nicely. Drystone lets air under your house, making it colder in winter but keeping it drier. I don't advise, however, letting your toddler stick his fingers into those inviting, dark recesses after large, curiously patterned "wiggle worms."

Foundation Tips

In all, foundation work is singularly unrewarding. You spend days bent over, digging, prying out rocks, pouring concrete, laying stone. And not much of this is visible progress. It seems to be a fact that builders, as well as those who inspect and criticize their work, like to see more happening faster. So you might want a minimal foundation, either drystone or footing-and-masonry just at the corners. You can always fill in later, you know.

But whatever you do, make it solid. Cover a lot of area with the foundation to resist settling. Get it high enough to discourage termites and rot, and even grass fires. You'll also appreciate the added crawl space when doing plumbing or electric wiring under there. Building code requirements of at least 18 inches of height are just barely enough.

And make it look right. Concrete blocks are cheap and fast, but ugly. And by the time you've bought them, laid them, then veneered over them, they're not cheap anymore. Solid stone can be the cheapest or most expensive, depending on whether you've had to buy the stone and pay a mason or collected it free and not paid yourself. Count on from $20 to $50 a square surface foot, including all materials, for stonework you hire done. Generally, tighter mortar joints cost more. Inspect prospective masons' work before you hire them.

Drudgery though it may seem, there's still a bit of a thrill in turning the first shovelful of foundation earth. You're actually on your way now; your log house is out of your head and on its way up. Enjoy.

FOUNDATION AND BASEMENT CHECKLIST

☐ Build a longer, wider stone foundation on softer ground.

☐ Plan for settling unless you build on solid rock.

☐ Have your footing ditch inspected before proceeding further.

☐ Check into the possibility of a basement if your site is a hillside.

☐ Seal basement walls with masonry coating and tar.

☐ Install plumbing pipes before slab is poured.

☐ Bring the foundation up at least a foot aboveground before laying the first log.

☐ Set the first logs and joist ends on metal flashing.

☐ Have the footing poured and let the concrete cure at least two days.

☐ Haul in more foundation stones than you think you'll need.

☐ Make the foundation solid, wide, and high.

CHAPTER EIGHT
Hewing, Notching, Log Raising

HERE IS WHERE your hewn-log house really becomes what it is. The design, foundation, roof, and additions can be similar to those in other houses, but those massive tiers of logs put your house in its own natural light.

Most American cabin logs were hewn, and there are lots of good reasons for this. No matter how straight logs are, natural taper will make fitting corners and window and door facings a pain. Hewing to a common thickness makes notching and facing easier, cuts down the work of stripping bark, and gives a flat surface inside for wood trim and hanging things like paintings and outside for rain to sheet off. It also cuts weight, quite a factor if just one or two of you are handling those logs.

Traditionally, a hewn-log house was also a status symbol because transient hunters and the poorest of settlers threw together huts of round logs and mud. Barns were built this way, and what self-respecting frontier woman wanted to live in a barn? Besides, hewn logs were the style both back East and up North. Heavy timbers were used in the New England clapboard houses and the English half-timbered structures, as well as in Scandinavia and central Europe. Our forebears came, and they brought their traditions.

Even if you restore a house on-site or move a house to your site, you will still need to know woods and the skills necessary to replace bad logs. Oak has been the favorite for hewn logs in the mid-South and Midwest, with pine most common in the East and Southeast, and rare examples in walnut and cedar. I delight in working with cedar, but it's hard to acquire. Our Mis-souri house was about half oak, with cedar having been added at some pre-1890 rebuild date, and more cedar when we restored and added to it. Chestnut and yellow pine were favored farther to the east. Whatever wood was used, the best builders (not all of them) almost always stayed with the same kind throughout. For some reason, mixed woods generally go with looser craftsmanship.

Chestnut was among the favorite woods of the pioneers east of the Mississippi. It was light, rot resistant, grew fast, and, with its distinctive grain, was often substituted for and mixed with oak. The chestnut blight destroyed this wonderful wood, but foresters are hybridizing Asian blight-resistant trees with the

Using a crosscut saw to trim a log.

ABOVE Beginning to hew a pine log. The scoring has been done with a regular axe. The right-handed broadaxe is being used to slice off the "juggles" — sections between the scoring — for a smooth, flat surface inside and out. While the top and bottom of the log could also be hewed, in this case the top and bottom will remain rounded.

UPPER RIGHT John is making the final pass on a hewn oak log during a log workshop in Bethel, Missouri. Chips and wood pieces are laid under the log to protect the axe. His is a right-handed swing.

LOWER RIGHT Most of the logs in a corner had to be replaced. The corners were sound, but too short. A new log was lap-jointed to the corner piece. Willie Lehmann is adzing the lap-jointed log to thin it for a match to the existing corner log.

few remaining American trees to try to bring back this treasure.

Avoid woods that rot easily, and those that are hard to work. Sycamore, elm, gum, and hickory are heavy and rot quickly, although they're generally available. Oak is not easy to work, compared with cedar or pine, but is long lasting, tough, and generally available. White oak has always been the favorite, but it is also in high demand today. Red oak (water oak, pin oak, whatever your local variety is called) works well with a broadaxe or adze, being inclined to split off in chunks easily. It doesn't last as long in the weather as white oak, chestnut oak, or post oak, but is generally easier to acquire. Post oak is tough and hard to work. Try to avoid it.

Hewing the Log

Hew your logs green. They're more inclined to crack open that way with shrinkage, but hacking a seasoned oak log is punishment. Folklore tells us that settlers hewed logs in the "M" months, March and May. Don't worry about the cracks; they're too narrow and shallow to matter much. But cut the trees when the sap is down, in November to February, to reduce this checking and lessen seasoning time.

Hew two sides, leaving the other two for extended height when laid up. Most houses were built this way, although some were of logs hewn on all four sides. In the Batesville, Arkansas, area there are many square-notched houses with all the logs hewn on four sides, to a common dimension. Some finely built log houses in the East, with four hewn faces, reflected the builders' European and Scandinavian backgrounds.

I prefer the broadaxe over the adze, which is more of a finishing tool. As far as I know, nobody manufactures these axes now, except for some rare and expensive hand-forged collector's items. They can still be had for a price at junk shops and antique sales. The axe head can be handled for right or left use, and these handles *are* different.

To hew, you stand alongside the log with the flat side of the axe to the wood, and it's on the same side with you, very close to your toes. Your knuckles also take a beating unless the axe handle is bent away from the log. So, with the bend away from the flat side, that means a different handle if you're left- or right-handed.

Novices usually devise their own variations of the basic straight-down swing, but anyone who does much axework soon returns to it. Years ago I thought I'd discovered something by using a right-handed axe at a 45-degree angle, hewing on the other side of the log with my left-handed swing. I soon abandoned that strenuous game, and I suggest you forget such variations. The heavy axe is more efficient used straight down with gravity working for you instead of against you.

A good man often hewed a dozen logs a day using the old method, and lived. I hew two or three, then find something else to do for a while. A good day's work for tie-hackers was also about a dozen a day, hewed on four sides but only eight feet long. I hear tales of hackers turning out 20 ties a day, but I would have to see it myself.

Craftsman Wilson McIvor helped us adze the faces of these huge sawn poplar logs for a new log house near Nashville, Tennessee. These logs were as wide as 34 inches.

First, score-hack the log down one side, every six inches or so with an axe. You can stand on the log for this, or chop from alongside, avoiding your kneecap.

Some hewers snap a chalk line along the log as a guide and hew to the line after they've scored to it first. Most logs aren't straight enough to make this easy, so other hewers sort of chop a line in the bark. I just eyeball it and make a second pass to take out humps. You'll need two passes at the thick butt end anyway.

You can also chainsaw these scores to a snapped chalk line. Master craftsman Peter Gott, in North Carolina does this neatly and efficiently. I have always eyed my logs, and chopped more than sawed. Then there are a couple of ways to go. One calls for splitting out the chunks, or "juggles," with the plain axe (poll axe, felling axe, "choppin" axe) then making a pass with the broadaxe. I score deeply, making a notch if necessary, then slice off the juggles with the broadaxe.

Either way, you eventually assume the broadaxe stance, swing straight down to hit the log at about 45 degrees, and cut off everything that doesn't look like a hewn log. Lay tough boards under, to keep your axe out of the rocks (after a couple of logs, there'll be plenty of juggles to pad the ground underneath). Early hewers set the log up high on blocks and hewed almost straight down with the broadaxe, which had a short (two-foot) handle for precision. I like the later three-foot handles, which I usually make myself, because you get more swing and take off more wood at a stroke. The longer handle does require more accuracy.

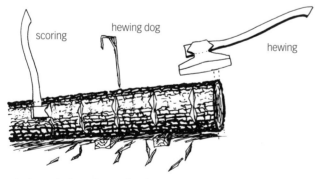

scoring

hewing dog

hewing

The log to be hewed was often held in place by a log hewing dog. First it was scored (notched), sometimes to a snapped chalk line, then sections (juggles) were sliced off with the broadaxe.

Practice splitting match stems with the axe for a few years and you get the idea.

A log rolls around a lot while it's being hewed, and you should stabilize it. I sometimes hew the log where it falls, leaving the treetop connected and maybe even part of the stump to steady it until I'm through. If you can't do this, flatten two places on the underside of the log and roll it onto crossways timbers. Or get a pair of hewing dogs, which were iron stays driven into the log and the cross timbers to hold things steady.

You may have read about hewing from the top of the log down, which is fine in theory, but knots tend to head uphill and are easier to slice through from the butt end. Although the swell at the base is easier to hew from the top, I like to cut this off anyway. Switching means rolling the log over. Being able to chop either right- or left-handed, I keep two axes handy and use both. That way I torture new muscles.

There is a mystique about converting tree trunks into hewn timbers that is quite compelling. It is Herculean labor, but it will nevertheless remain one of the most satisfying experiences of your log house building — in retrospect, that is. With that heavy, heavy axe, you are shaping a second life for the tree you've felled, feeling the steel bite into living wood, to ringing echoes of the hill country past.

You're Pulling My Leg

Let me here digress to try to clear up what I regard as misinformation about hewn log walls. I have read and heard of the practice of constructing round-log cabins, then hewing down the walls *in situ*. Nancy McDonough, in her book *Garden Sass*, quotes an old-timer from the Ouachitas as being familiar with this practice. D. A. Hutslar mentions "scutched" logs as those superficially hacked after the house was up. This was sometimes done to a thicker log when furring strips were nailed up later for siding to go on.

I've used the broadaxe for many years, and have probably hewn more logs than the average pioneer in his entire life, yet I wouldn't attempt to hew logs in a standing wall. My brother is an artist with a broadaxe, and can shave a ⅛-inch layer end to end, and he wouldn't try it either. I know dozens of veteran hackers who couldn't do it, and never heard of its being done.

These short log sections were left when adjoining windows were sawed out of the original cabin. When building new, very few logs need be full length. These spanners above windows and doors are full length and carry the ceiling joists.

Working the lower half-dovetail with the adze. The log is upside down.

Apprentice Willie Lehmann and my son, Charlie, lay out a top half-dovetailed notch at a demonstration in the National Building Museum in Washington, D.C. The slope can either be taken from the notch of the log above, as here, or laid out with a tapered template and level.

I have visited thousands of hewn-log cabins and have found no evidence, from a hewer's experience, that any of them were so shaped. It's the sort of thing that may have been speculated upon, because we know that logs were rolled up skids at log-raisings and maybe we assume they were still round at this point. To get logs in place, I've used skids, I've rolled them, but usually I've slid them up; in every case they were hewn sensibly, on the ground.

Of course, it would have been possible, with complicated scaffolding to stand on and a wanton disregard for digital extremities, but it would also have been somewhat self-defeating given the added weight and difficulty of notching, and so on. No, except for rare freaks, round-log cabins stayed round, and hewn ones were shaped on or near the earth's surface.

Log Lengths

When figuring log length in the woods, start with your outside dimensions, and allow about two inches more at each end to extend past the notch. Most early hewn-log houses had the log ends trimmed flush, and you may want to do this later, but allow the extra length for an error margin. So an 18-foot inside dimension requires at least a 19-foot-4-inch log, allowing for a 6-inch wall thickness. Before you begin, figure the number and placement of windows, doors, and fireplaces, so you can use short log lengths. Although it

seems faster and simpler to cut all the logs full length and then chainsaw out the openings, it's not. I've done houses that way, and the pioneers always did (*sans* chain saw), but there are good reasons not to do it that way when you're building.

First of all, you'll need very few full-length logs — just enough to bind it all together. And it's easier to handle short ones. Long, straight logs will be at a premium, no matter where you are, and meandering tree trunks will afford lots of three-foot and four-foot lengths, perfectly usable. Also, full-length logs must be notched exactly on both ends so the log won't teeter, and this takes several tries. A segment, from a corner to a window, can be a fraction off without threatening the structure, and no one will be able to tell.

So save time, cut to length, and hew just what you need. A 19-foot 4-inch log, with a 3-foot door and 2-foot window cut-out, becomes 3 short lengths, maybe 6 feet 2 inches, 4 feet, and 4 feet 2 inches. And the notching can be faster and a bit less demanding than on the necessary full-length logs. The full-length ones can also weigh as much as 600 pounds after hewing, green, so handle as few as possible.

To keep the loose ends of short logs in line, you can block the spaces between logs with 2×4 end pieces or scraps to maintain a level. Spike a scab board up on the outside or inside surface of the log, temporarily, near where door and window facings will go. These vertical boards come off later, so they can also be scrap.

FAR LEFT **Cautious use of a chain saw makes quick work out of notching a tough oak log. My brother John is a master with this sometimes dangerous tool. Hand tools offer more time for control. Safety goggles are a good idea in any construction project.**

LEFT **Setting a half-dovetailed notch into place.**

Decorative diamond notches on a 1763 silversmith's cabin in Leesburg, Virginia. The notch is actually a functional V-notch with the ends dressed to the diamond shape for appearance.

The V-notch in these oak logs was typical along the eastern seaboard. Later in this area and farther west, the half-dovetail notch became common.

This is the half-dovetail notch or chamfer and notch found on most American hewn-log cabins. This example in oak is from the Ozark Mountain area of Arkansas.

Notching

You'll read about V-notching, saddle notching (round logs), full-dovetail notching, tenon (square) notching, half-dovetail (also called dovetail) notching, and probably some others. The traditional log builder used the chamfer and notch, or half-dovetail notch, which is simply a vertical cut on the underside of the log to a depth that leaves the desired space for chinking, plus an angle cut toward the end. This is the notch. The chamfer is the sloping cut on top of the log that lets the next cross-log notch fit over it.

Our own Missouri cabin had been scrambled at least once before, in rebuilding, and several of the logs and their notchings were upside down. And there were some full dovetails among the halves. We decided to leave them that way instead of reshaping the brittle old wood. The cedar logs I added were all half-dovetail, right-side up.

This half-dovetail notch pulls the corners tighter together both ways with the weight from above, which some other notches do not. The V-notch and compound-angle dovetail are the other two best choices. These also allow rainwater to drain out, which helps prevent rotting. Wide eaves help here too, a non-pioneer practice that keeps making more sense as time goes on.

Notching should follow a pattern based on a set angle, and a set amount of wood left, not wood taken out. So notch a swelled butt end deeper than the thinner top end, but keep the same angle. Cut yourself a pattern to use on the hewn logs. We use a two-inch rise in a six-inch thickness. Make sure the pattern is laid pointing straight to the other end of the log, or curves, kinks, and flaring ends will throw you. This is where a snapped chalk line down the centerline of the log can help you. It's possible to use a square and template to get the notch nearly perfect. This takes about as much

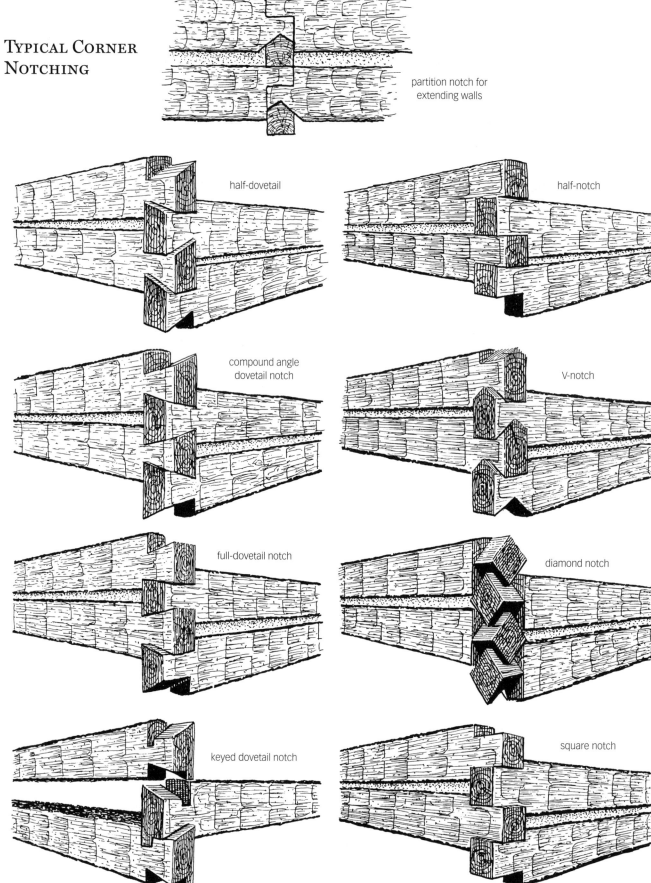

Typical Corner Notching

partition notch for
extending walls

half-dovetail

half-notch

compound angle
dovetail notch

V-notch

full-dovetail notch

diamond notch

keyed dovetail notch

square notch

time as estimating the angle, then adjusting after the first trial, which you usually have to do anyway.

The same angle isn't really essential. A friend who helped me with one cabin always worked the chamfer about where he thought it should be, then used an adjustable T-bevel to take the same angle to the next notch. I do it all by eye.

I use a crosscut saw to make the vertical cut, but my brother John, who has a steadier hand, likes the chain saw. Remember to allow for the wide cut the chain makes. Then we both use an axe or adze on the angle cuts. You may not want to risk these long-handled tools while perched up in the sky, which is where we usually work the chamfers. A large chisel will do nicely, but if you axe things, I suggest steel-toed boots.

You will almost certainly cut out too much wood eventually, so go light till you get the hang of it. Too wide a chinking gap means cutting the notch or chamfer deeper. Don't do this unless you need a wide gap to keep the log level. More about that later.

If you do take out too much, and the log lies against the one under it without closing the notch, you can hew away part of the bottom edge or the top of the one under it. Or save this log for later use. Or, you can slip a piece of board into the notch as a spacer. This is not done in the best craftsmen's families, I realize, but it's sometimes necessary when restoring a house that has parts of the notches split off.

In notching, keep an eye on the other corner, so that your logs will be the same height as you work your way up. You'll span the opening at its top, and this spanner log should be reasonably level. So measure each corner, from the sill up, each course or so. Stay within a couple of inches.

Some years back an aspiring log builder I knew got the job of putting up a house I had consulted on. I visited the site one day to discover that he'd built an addition log wall, not paying attention to the relative heights of the logs on each side of a doorway.

This man was a huge, high-energy individual whose approach to building reminded me of a charging bull. If a dovetail notch wasn't a tight fit, he'd take swipes through it with a giant chain saw that resembled a locomotive until it was different.

I pointed out to him that, at the top, his spanner log wasn't going to span because one side of the door

was a foot higher than the other. He'd "adjusted" too many notches on one side and that let the wall down.

He caught on fast. With a swipe of a mighty paw, he sent log sections flying, and without a pause began carving out replacements with that saw — which takes us to keeping the whole thing level. Only theorists

This is a square notch on an 1890s cabin log. Square notches were used in later cabin construction, but did not provide the interlocking advantages of the dovetail.

Willie Lehmann uses the slick to smooth a half-dovetail notch for a better log fit. The slick is pushed with a side motion to slice the wood.

assume that logs are straight and of constant diameter. Lay the courses of logs alternately; place the first log with its big end one way and then lay the next one over it with its big end the other way. And alternate opposite wall logs. The best plan is to lay each course with the big ends joining little ends all the way around. The next course should join little ends with big ends, corner for corner. This will make notching much easier, and make it easier to keep all the logs more nearly level. I inspect over 100 log houses a year, and have seen only two that were built with all big ends together. One was an elaborate diamond-notch design from southern Virginia, the other was Muskrat Murphy's Arkansas cabin. They had very wide chinking cracks at one end and quite narrow ones at the other.

Use a layer of flashing between the foundation and first logs. We use copper, which lasts, and becomes virtually invisible. This keeps moisture from wicking up, and stops termites cold. Don't let it extend out or it'll catch rainwater and rot the log anyway.

In a log house, your concern with levels is limited mostly to the floor, ceiling joists, if any, and top plates. Window and door sills can be cut to a level; so can mortises for floor and ceiling joists and rafters. But it's easier and less disturbing visually if the logs involved are approximately horizontal. If you notice one corner is high, notch the next log deeper, even if you have to shave off some of the butt end to let it sit down on the chamfer. Leave two to four inches for chinking; more means too much mortar and looks sloppy, and much less makes it hard to get the stuff in.

Sawing the vertical cut for the half-dovetail notch at a log workshop. I require all students to start with hand tools. We switch to power later. Hand tools work well when sharp.

Alternate small end with large end, as shown in this exaggerated drawing, to keep log courses level. Check corner heights often and stay within two inches of level, if possible. Adjust with deeper notches when necessary.

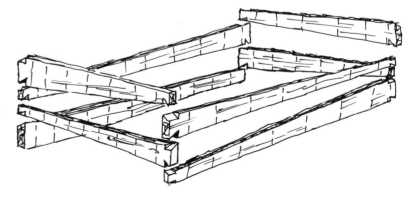

Floor Joists

I like to lay my floor joists mortised into the first logs, or sills, at a height well below that of the second (end) logs, so that the third pair won't have to be notched for the tops of the joists to fit into them.

Shall we do that one again? It's easy to get in trouble when trying to figure final floor height. So here's a good rule: Lay your sills as front-and-back logs, and mortise them for the joists so that a maximum of two inches of the joist is above the sills. This way your next pair of front-and-back logs will clear or just touch the sill tops without more mortising or trimming, and still fit into the end log dovetail chamfers.

Joists can be sawn timbers, logs, or beams. I like 16-inch centers, and support in the middle for over 18-foot spans. A stone wall, or a heavy sleeper or girder, as this support is sometimes called, will brace the middle. Wider centers will hold the weight, but will let the floor shake and creak. For hewn- or round-log joists I like the same chamfer and notch into the sills as at the corners of the house. But for thinner timbers, a neat mortise is best, tight enough to keep the joist steady.

If you use two-inch stock, for instance, these mortises will make fewer crisscross braces between joists necessary. The building codes will require 16-inch o.c. joists, of a specific size. Usually 2×10s will span up to 18 feet without a sleeper support, but blocking or crisscross bracing every 8 feet is required. Heavier (two-inch) flooring will let you space floor joists wider, but it's expensive. Most codes require a minimum of one inch of total floor thickness, including subfloor.

Take care notching or mortising in your joists; you'll have to live with that sloping or wavy floor for a long time. Keep the tops, where the flooring goes, level. To avoid taking out too much wood from your sills, you can cut an eight-inch joist down to four inches at the ends without noticeable weakening; it's only in the spans that they tend to sag. But building codes allow only one quarter of the total to be cut out here (two inches of the eight-inch joist).

I don't like to lay the floor till I have a roof on, because rain and sun will warp it, shrink it, and generally play hell with it. I usually lay some two-inch boards or exterior plywood across the joists to stand on while building.

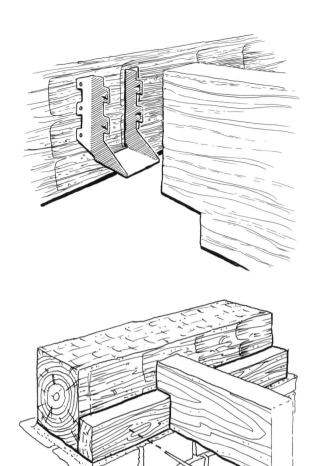

You can use joist hangers instead of mortising the joists into the sills. These are galvanized "pockets" nailed with special, heavy nails into the faces of the sills. Or you can use a nailed or lag-bolted ledger strip across the bottom of the sill face to set the joists on. Or set them on metal-flashed or pressure-treated boards set on the foundation itself.

These floor joists are mortised into the oversize sill log for stability. Floor joists can also be set on continuous foundation shoulders or attached to the sills with joist hangers.

This is an example of an extended log wall where a three-way notch is used to stabilize the end-to-end log with an intersecting log wall. Not all the wall logs are end to end here, between windows. In this case, the stub wall is a minimal two-foot wall because the owner wanted as much open space as possible.

Door and Window Bucks

Spike door, window, and fireplace bucks directly into the ends of the logs, but not until the logs have seasoned and settled. Make a clean cut here. I leave the log ends long until they're up, then make one long cut. I like using the crosscut saw because I can control it better, but a chain saw is faster. Only the greatest need for haste drives me to use one.

One-inch oak or 2-inch pine is heavy enough for these facings, and a 20-penny common nail is large enough; use four nails per log. Don't use box nails in oak. Those slender fasteners were designed for soft pine, spruce, or fir. Even with common nails you may have to drill an occasional hole, especially if your logs are seasoned. I prefer the modern counterpart of the old cut nail, square, now hardened for masonry, which will go through anything. It's expensive, and you need to drive its wedge shape with the grain, not across it, to avoid splitting. Our forefathers bored and pegged these facings with trunnels, the folk derivation of *treenail*.

The facings (backs) hold the ends of the wall logs in line, and should be mortised into and fastened to the top spanner overhead and window or door sill log beneath. After they're in place, you can pry off the temporary strips (scabs) you nailed up as the logs went into place. It seems it would be easier to start with the facings themselves, but it's easier to trim the log ends evenly when they're all up. And the logs need to settle; allowing space for this has never worked for me. The pioneers had an abundance of timber and built solid log walls. Then they used the crosscut saw for the window and door cuts.

Preparing for Chinking and Electric and Plumbing

A time-saver before you raise the logs is one I learned from Menno Kinsinger of Virginia's Shenandoah Valley. Cut a chain-saw groove an inch deep about 1½ inches in from the faces of the log on the underside. This is for the sloped chinking wire mesh to slip into later, saving tricky fitting and nailing.

We also drill holes near door openings for electric wire before we raise the logs. We use a one-inch bit to drill up to wall switch height, then continue on up through the top plate in at least one place. (See chapter 11 for more on this process.)

Raising Logs

Getting the logs up onto the walls is a major undertaking. Two to four men can wrestle 16- to 20-foot hewn logs into place if the logs are under, say, 12 inches in diameter. But these are hard to lift to any height. Long skids were usually the answer for the early settlers, with the logs skidded up by hand or cross-hauled by oxen or horses. Although it's slow, a block and tackle will do the job. Skids or a vertical lift from a tripod or boom provides the base for action.

The device I used for years was a full oak 2×4 A-frame (gin poles) pivoted from the front bumper of my Land Rover and braced with a guy cable over the rig down to the trailer hitch. I ran the winch cable up over a sheave at the peak and down to the log. Lacking a winch, a ratchet hoist or "come-along" will do it by hand. Either way, the log should be lifted at two

points for balance. Once the rig is in place, it's a good idea to support each end of the bumper with a block, to take the weight off the springs and spindles. I've since replaced this with a 1951 Dodge Power Wagon with winch, and braced two-inch steel pipe for the gin poles.

On very tall houses, we sometimes found my gin poles weren't tall enough. That happened on our Virginia house, and I was resigned to cutting and welding in extensions. Then, a woman who worked for us suggested driving the Land Rover up onto my flatbed truck to make it taller. It took 20 minutes, and it worked.

At a house-raising, four men with ropes can haul each log up from the top while others push from below. Or you can use skid poles with the ropes.

I copied Peter Gott's hand lift a few years ago, and it works well on relatively small logs, say up to 20 feet and maybe 12-inch faces. It's oak, cut 1½ inches thick and braced often to keep the weight down. I lay a subfloor to use the lift so it can be moved around from wall to wall. Peter uses big casters on his, which is heavier with more lifting capacity. I've lifted 400 pounds with mine, with the lightweight boat trailer winch on it. This device lets you lift logs from the inside to each wall. Often a site won't give you access to all four walls on the outside, so this is an advantage.

If you have room to maneuver and the money for it, you may hire a knuckle boom. These are usually mounted on 10-wheel trucks and have a reach of up to 30 feet or so. They're logging machinery, and a good operator can take down a house with one in a hurry. He can set it up quickly, too, but only if everything is prenotched and ready. Fitting as you go is too slow to make a knuckle boom efficient.

Check your log wall for vertical alignment as you go. Particularly around windows and doors, you'll have to

Cutting a groove for the chinking wire mesh that will hold the masonry mortar. Two grooves are cut on the underside of the log (above left) with a chain saw, which greatly speeds up the chinking process, as well as making the chinking look straighter and neater.

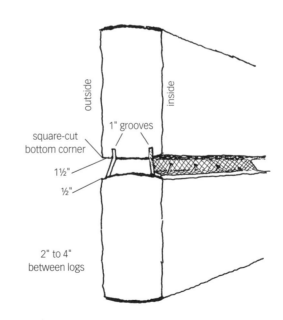

Installing the wire mesh for mortar chinking is simplified by cutting these grooves with a chain saw on the underside of each log. Log surfaces sometimes need to be prepared for the chinking process. The wire must slope outward so the finished chinking will shed water.

Raising Logs

Regardless of how logs are lifted, it must be done safely. The size of the cabin, the terrain upon which it will sit, and your budget will determine what method you will use.
(A) The Land Rover was elevated by driving it up onto the bed of a truck; this creative use of gin poles to lift the logs was born of necessity, but is not a recommended method. (B) In pioneer times, cabins were erected by as many hands as the owner could corral; here, workshop participants heft logs. (C) A knuckle boom works well also. (D, E) Whether it uses a strap or log tongs, a crane is very efficient, especially if it can be parked in one spot and has a reach long enough to put all the logs up from that spot.

watch a tendency to slope inward or outward, as many old houses do. That's of no great consequence until you find your door won't open without swinging into the floor. Use a level, and make sure all your full-length logs are the same length from notch to notch. Some delicate adjusting for alignment may have to be done from time to time. I use a sledgehammer.

Ceiling Joists and Ceilings

At last, you're up to ceiling height. You should have at least one log spanner above windows and doors, with the facings attached after seasoning. If you cut deeply into this log to get your door or window height, go up another course before laying the ceiling joists. You'll mortise these joists into the log, and you must be sure there's enough wood left to support the weight.

You may not even want a ceiling or a loft, but the joists are still a good idea; run them from front to back, at right angles to the ridgeline. They are the truss chords that tie the rafters together; they keep your ridge from sagging and the rafters from pushing the top logs apart. There is another way around this, however, which we'll talk about when we get to the roof.

Let's assume that you do want a ceiling, and a floor for the loft. These can be two layers, on top and bottom of the joists, with insulation between. But because you'll be using the loft, let's not insulate it from the downstairs heat. And let's keep the ceiling beams visible. Massive, hewn joists support better, and look better by giving a reassuring feel of strength and shelter. I hew them with the broadaxe, then adze for additional smoothness. Fancy houses in the old days sported beaded joists, worked with a hand plane.

A 6×8 joist is both aesthetically pleasing and strong. But for a house deeper than say 16 feet, building codes will require something deeper, like a 6×10 or 6×12. Lay the joists a minimum of 4 feet apart, or 3 of them in a 16-foot house interior, 4 in a 20-foot interior. They should be half-dovetailed into the top front and back logs, so that they tie these walls together. Sometimes they were simply mortised, but this gave no bracing against outward thrust unless they were also pegged. Dress the top surfaces of these logs and joists carefully, as you did your floor joists, because you'll be nailing flooring on top of them.

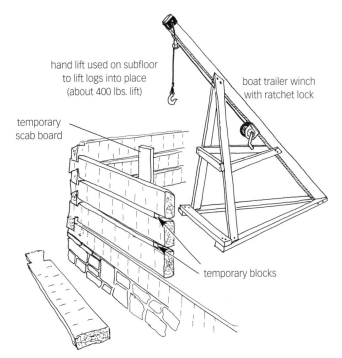

hand lift used on subfloor to lift logs into place (about 400 lbs. lift)

boat trailer winch with ratchet lock

temporary scab board

temporary blocks

This hoist is the invention of Peter Gott, master log builder from Marshall, North Carolina. Peter says he is glad to share this device with other log builders, because we all borrow building techniques from each other whether or not we realize it. It's used on a subfloor to reach over each wall to lift outside logs from the inside.

This simple lifting device utilizes a pair of braced 2x4s (gin poles) from the bumper of a pickup truck. Lifting is by a ratchet hoist or come-along.

Enough people can carry logs and put them up by hand, as in this scene from Murphy's log raising. The stone-corner foundation was filled in after the cabin was up.

Probably the best flooring-ceiling combination is two-inch-thick tongue-in-groove lumber, which is quite expensive. A one-inch subfloor, with a second layer over, is also acceptable. Tongue-in-groove lumber helps each board support its neighbor in the spans. If non-grooved lumber is used, you'll need closer joist spacing, traditionally two feet.

For our Missouri house, we were able to obtain, albeit at great expense, full 16-foot lengths of 2-inch tongue-and-groove lumber, which spanned the entire inside, leaving a neat inch extending into the chinking space between logs at each end. A word of caution, however: Try to lay this ceiling-floor in very dry weather. Even kiln-dried lumber can shrink enough to pull apart and leave cracks when winter heat is applied. Now I order this shrink-wrapped in plastic at the mill to keep it from swelling with moisture.

Again, don't lay this ceiling until you have the roof on, unless you're one of those optimists who can safely plan to have it on before the next rain. It would indeed be easier to nail the ends of the boards to the side logs, if nothing were above them, but you can slant the nails fairly well. Or use a nailer strip against the logs for the flooring ends.

Loft Space

Now, assuming you want to make full use of your loft space, you'll need to lay at least three more courses of logs above the ceiling level. Notch out the next front-and-back logs to fit over the joist ends and go ahead. This may take awhile, because these notches must match the end dovetails. Obviously, additional logs above these joists just wreck their function as chords in the rafter-truss game, but do it anyway. As I said, we'll take care of that problem later.

Once, early in my career, I wanted a ceiling at a height that meant running the joists parallel to the ridge. There seemed no real reason not to do so. (The joists had no connection to the roof.) My brother John and I were building this cabin some distance from home, and left the job for the weekend. We were both convinced that the joists should be at right angles to the ridge, but, because they would give no support to the rafters, neither of us could reason why. We puzzled over it all weekend.

The alternative was another pair of logs and more ceiling height, which we really didn't want. We'd already perched the cabin on 4-foot foundation piers, and having only 16-foot-square outside dimensions, it was already almost too tall. Besides, I knew I'd seen houses built with the joists parallel to the ridgepole.

So we laid the ceiling joists parallel to the ridge. And it wasn't until we put in the stairs that we saw why this is a mistake. Stairs usually run along a wall, and you bump the roof if you go up parallel to the ridge. You could move to another wall and cut a couple of the joists and support them from below, but that's messy. In this case, we'd planned a disappearing ladder stair anyway, so we put it in the middle of the room between two joists and left everything as it was.

Stairs do take up a lot of your precious room, and many old houses used narrow, steep stairs or ladders nailed to the wall, or, later, disappearing stairs. With the joists laid at right angles, disappearing stairs can be placed neatly along the side wall opposite the fireplace, like regular stairs. There is more about getting upstairs in the chapter on lofts.

Above the joists, lay as many courses as you like to help get headroom. But remember, the top course must not be cut into to allow for windows. These logs need their full strength, as you'll see when we talk about roofs.

Overhead joist headed off with a mortise-and-tenon joint for a stairway. The upstairs flooring is two-inch tongue-in-groove to span the wide joist spacing. Joists are recycled 3×12 heart pine.

This 32-foot-long summer beam in the Wintergreen house supported the joists for their 26-foot span. The summer beam was 18 inches in height. After it was set in place with a knuckle boom, I adzed it, nine feet in the air. The photo shot from the top of the beams shows the complex way in which ceiling joists were notched, then braces bolted into the summer beam.

Top Plates

Finish up with the front-and-back plate logs, or you can top these with a pair of end logs. These will be the chords for the end rafter trusses, and will carry a lot of strain until the gable ends are nailed in.

Try to get all the logs level at the top; you won't be able to get a tight fit to the roof front and back anyway without chinking, caulking, or boards fitted in, but it makes things easier if they're level. Do drill and peg the plate logs in place, using heavy hardwood or steel pins. You will probably want to extend the plates a foot or more at each end to carry a pair of eave or trim rafters. It's a good idea.

If you hew or saw a heavy plate on four sides in the old tradition, you may terminate the rafters here, or notch a bird's mouth for an eave or overhang.

If you've put your logs up green, you can still go ahead and roof to keep them drier. Just don't nail in the facings at window and door cutouts until shrinkage and settling are complete. If the logs are completely seasoned, you may go ahead with all the finishing of the house. Either way, it's not a good idea to leave the logs up unroofed because things have a way of distracting you. Too often I see a rotting stack of logs that somehow never got roofed. Chinking can wait; windows and doors can wait too. But go ahead and roof it now. Don't let anything happen to all that labor and investment in creative energy.

HEWING, NOTCHING, LOG-RAISING CHECKLIST

- ☐ Avoid woods that rot easily and those that are hard to work.
- ☐ Hew logs while they are green.
- ☐ Cut trees when sap is down — November to February.
- ☐ Figure log length according to outside dimensions, allowing two inches to extend past the notch.
- ☐ Hew only what you need.
- ☐ Notch to follow a pattern based on a set angle and a set amount of wood left.
- ☐ Keep the logs level.
- ☐ Trim log ends at door and windows after they have seasoned and settled.
- ☐ Check log walls for vertical alignment as you raise them.
- ☐ Lay floor and ceiling in the driest weather.
- ☐ Roof as soon as your logs are in place.

CHAPTER NINE

Roofing

ROOFING YOUR HOUSE gives it the final shape it will wear, and of course lets you start keeping things dry inside. Any number of styles and procedures are acceptable, but I'll outline the one most used by most settlers after sawn lumber became available, which incidentally is also the simplest and most practical.

A 45-degree (12/12) pitch was common because it was easy to figure, was steep enough to shed rain and snow but not too steep to work on, and gave headroom in the loft. The kids were stowed away up here, where wind whistled between the shakes in winter and only a little warmth found its way up from below.

Gable ends were sometimes just more logs, tapered and shortened to the slope with log purlins lengthwise and no rafters. This was done throughout the country, mostly before sawn lumber became available for the gables. The classic but unrestored Henderson cabin, near Hemmed-in-Hollow in north Arkansas, was built this way. This method requires more logs and results in a very heavy roof.

Ridgepoles and Rafters

A ridgepole was rarely used because the slats that the shakes were nailed to served the same bracing purpose. The rafters were cut to fit together at the peak, and were braced temporarily until these slats could be nailed down. Earlier cabins had mortise-and-tenon joints or were half-lapped and pegged at the peak.

Before you begin with the rafters, bore an auger hole down and through the top plates, right in the notches, well into the top end logs (I use a two-inch auger or expandable wood bit). Now drive in a well-

seasoned, tight-fitting oak, locust, or hickory peg and cut it off flush; or you can bore a small hole and use a steel pin, such as ¾-inch rebar. This is to hold the top course of logs against any outward thrusts. The settlers often used a very heavy, wide top plate front and back to hold the outward push of the rafters, and pegged the ends this way. But even a big beam will bow out as the roof weight settles, with no chords to hold the rafters together at the eaves. Of course, this

John and I peg pole rafters into the top plate.

is a problem only when you have a loft and log knee walls built up above the ceiling joists. Otherwise, as we said before, the joists act as chords in the rafter trusses.

Cut your rafters the length of the hypotenuse of the right triangle formed by the height of the peak and half the length of your end logs, plus your eave dimension. If you didn't follow that, let's suppose your cabin is 16 feet deep on the outside. A 45-degree roof would peak eight feet up, the same as the distance from the midpoint of the wall directly under the peak, to the plate. The high school formula tells us to square both eight-foot dimensions and add them to get 128. The square root of this is about 11 feet four inches, and you'll need an extra inch or two to be safe.

Any good carpenter can also lay out rafters with a framing square, using the pitch numbers (12/12 in this case). Each run of 12 inches is matched by a rise of 12 inches until half the house depth is reached, or the distance from outer wall to roof peak. The resultant rafter length is right, with a straight-down location for a bird's-mouth notch at the plate, if used. Any overhang is added.

Eaves were almost nonexistent on the old cabins, which is a significant reason so few have survived. I like one feet or two feet for eaves; more makes the cabin flop-eared and less lets the walls get wet. You don't need the extra overhang where a porch or lean-to will attach, because the rafters for these will probably tie in to the main ones. If you haven't decided on these additions yet, keep the eave. It's easier to cut rafters than to add.

Rafters were often as small as 2×4, which was too small, but they may be as large as you want. I like a rough-sawn or hewn 4×6 dimension, but you may want a deeper rafter for more overhead insulation. (We've talked about 2×10s for the nine inches of fiberglass insulation to meet building codes.) I have also used round cedar, pine, or oak poles. If worked by hand, hew or adze the top surface of the poles smooth for slats or decking.

Space the rafters 24 inches o.c., which is about all you'll want to span, no matter what your roof is decked with. Again, building codes will want 16- to 24-inch centers. Notch the rafters where they pass over the top plates, with a right-angle bird's-mouth cut at

A one-story cabin with hewn-beam roof trusses being raised by hand by our crew in Virginia. The open ceiling joists hold against the outward thrust of the rafters in one- and two-story houses, but one-and-a-half stories require additional bracing.

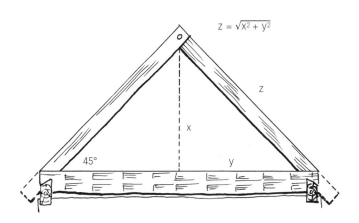

$$z = \sqrt{x^2 + y^2}$$

Rafter length can be determined for any roof by the formula above.

45 degrees. You'll find, unless you've been living right, that these plates are nowhere near level or straight, so cut mortises to the lowest level. If a plate log is rounded at the outside corner, mortise deeper to get to a square shoulder.

If you've hewn the plate on all four sides and terminate the rafters against it, make sure it's level. Of course, this gives you little or no eave, but was standard practice in log house construction. A minimal eave was provided when the plate was thicker than the logs below it, and extended outward, carried by longer end logs.

Assuming you're notching the rafters into the plate and extending them, use a level and chalk line on the face of the plate and mortise to it. Getting the rafters notched to give a uniform height is essential to avoid a roof that humps and swells like surf. It's easier, obviously, with rafters sawn or hewn to exact dimensions. With round poles, just notch a bit, try for fit, and notch some more. A layout string stretched from rafters set at each end will make this easier and more accurate.

I peg rafters right through the notch into the plate. A one-inch peg is enough, but you may also toenail it with a couple of 20-penny spikes. In a windstorm, a sudden low-pressure cell can cause your roof to lift off intact, if it is not secure at these points.

Here, the ideal peg is one with a bit of a head to it, with a hole tapered out in a countersink to hold it. A lag screw is probably better all around, but I've also used bolts driven into slightly smaller lead holes.

When measuring rafters for this plate notch, be sure to mark the apex of the notch. Almost every builder sooner or later starts the edge of the notch at the proper point, making the apex off a bit, then has assorted rafters that are an inch or two too long. It's the sort of mistake that's okay if you do it consistently. Otherwise it can cause your ridge to weave. Cut the bird's-mouth notch about two to four inches deep, just enough to help hold it in place until you peg it and to help with the weight.

Now, you'll need to cut the top rafter ends to a 45-degree angle to fit at the peak. No matter how careful you are with round poles, there'll still be variations, so I have used this method: Measure the total house depth, or the distance between each plate at the outside, at the point each pair of rafters goes. Lay the pair out with plate notches exactly that distance apart. Now cross the other ends at the measured peak point, and saw down through both at once. They have to fit that way. A log sitter is helpful here to hold things in line; with sawn rafters, you can get a 45-degree cut easily.

Now spike these cut angles together and reinforce with one or two 1×6 board plates, cut at a 45-degree angle, near the peak. That keeps the roof halves from splitting apart in a gale. It's also a good idea to nail a temporary brace across the rafters at plate-notch height or put in collar ties. This makes it easier to set the rafter pairs in place.

Traditionally, each rafter was half-notched or mortised at the peak to fit with its matching tenoned one. Then both were bored and pegged. This is a simple operation for a 45-degree roof, but more complicated when the pitch varies. The important thing is to hold the matched rafters together at the peak, against outward thrusts from low-pressure cells in windstorms or to protect the roof against falling trees.

You'll set each rafter pair, or truss, onto the log plate at the right spot or into the mortise you cut for it. Lay these out, dividing the plate length evenly for rafter spacing. Have the outside of the end pairs of rafters flush with the outsides of the log walls. If the plates are extended for a trim rafter pair, let these come at the plate ends, no matter what the spacing.

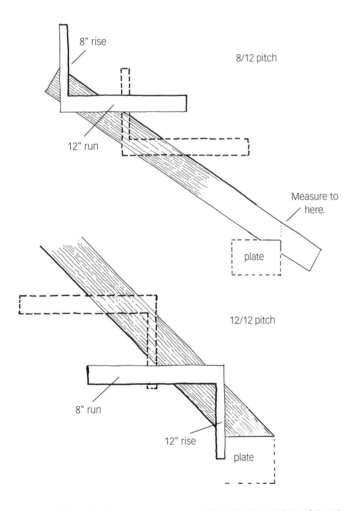

Using a framing square to lay out rafters. The rise-run legs of the triangle give the hypotenuse (rafter length) when the known center-to-outside wall is measured off.

Measure rafter bird's mouth from its apex to the peak. A common mistake is to start the cutout at the mark, throwing the roof off.

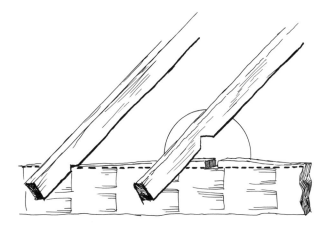

A chalk line on an uneven top plate will give the depth to notch to for rafter bird's mouths. Sometimes we use sawn beams for plates. The space between the rafters will later be filled with bird-stop boards nailed in place vertically. See drawing on page 109.

Then, having raised the first truss, level the whole thing vertically, nail a couple of angle braces temporarily, and peg to the plates. The first truss up is always cause for celebration, because now the vague outline of your total cabin begins to appear. Traditionally, a sprig of the tree type used for the cabin was affixed to this peak. Folklore tells us this was so a piece of the tree was the highest point of the structure, to appease God, who didn't want man-made things (Tower of Babel?) higher than his creations.

I usually do the two end trusses first, but you can start and finish wherever you please. The last one is always hard to jockey up, whether between others or at one end. Today, we often use a ridgepole to meet building codes, bracing it with opposing rafters at each end. Then we nail pairs opposite each other, working off scaffolding on the loft beams or stepladders set on boards over the beams.

If you have help, start at one end with stout pushes from a tacked-on pole and restraining pulls on a rope. Someone has to climb to the peak as soon as each truss is braced to untie the rope and/or pry loose the push pole. It's also handy to have a couple of stalwarts at each plate to keep bases from slipping. I tied the rafters loosely on our own Missouri house, and with the push pole nailed so that it could pivot as the truss went up, I was able to do it alone. After that experience I made it a rule that no one works on a roof alone.

It's a good idea to extend the top plates a foot or two beyond the end walls to carry the pairs of trim, or eave, rafters. If you don't do this, you'll have to use a look-

This diagonal brace carries roof weight to the gables, keeping the ridge level and allowing the roof weight to bear straight down. No outward thrust is exerted, even with no collar ties.

out to hold the eave overhang. This is a series of timbers mortised into at least the last two house rafters, to cantilever out and hold the trim rafters.

Dormers in Roof Framing

Dormer windows are a good way to let light into a 1½-story cabin's loft, since the logs in the knee wall should never be cut, for windows or anything else. Older dormers were quite narrow, three feet or less, with

The rafter system erected for a one-story cabin will have diagonal braces installed between the rafters to prevent outward thrust.

Roof and dormer framing for a one-and-a-half-story cabin shown from the interior. Bird-stop boards will be installed to close up the space.

peaked roofs to match the cabin's roof pitch. Dormers look good directly above the windows or door in the cabin walls.

One dormer by itself is out of place. Two or three are better visually. They also allow more floor space in the loft, usually being even with the outside walls.

Dormers are like little houses — not easy to build. Every carpenter I know has his own way of putting them together. When we do more than two on a house, we sometimes pre-build them in the shop, then set them up with a crane or boom.

On the downside, a dormer does let heat out because it has less insulation than the main roof does. It would be out of proportion to match the 10-inch roof rafter of the main roof by insulating the dormer as heavily as the main roof. We use 2×6 rafters and dense Styrofoam insulation for as much energy efficiency as we can get without making the dormer ugly.

Some people put shed dormers (flat roof) or eyebrow dormers (hump roof) on log cabins. They're invariably ugly and out of place.

Roof Bracing

Now for the permanent roof bracing, which serves a dual purpose if your house has a loft with logs above the ceiling height. Angle four braces, say 2×4, from each corner up to or near the center of the ridge. If you don't have a rafter truss at the center, go to the truss at each side.

Got that? I thought not. I've had trouble explaining this brace even to architects, unless I have a pencil and paper handy. It's an angle brace on the same plane as the roof slope, from each corner to the center of the peak. It can be spiked to each rafter on the way and notched into them. Or it can be in segments, between the rafters and flush with them. I like the latter better because it gives more headroom upstairs, allows for possible finishing inside the roof, and doesn't weaken the rafters.

Now, these braces eliminate end shifting, which is fairly obvious. You could also do this with braces angled from the ceiling joist centers up to each peak end. But this doesn't work as well, and you'd cut your loft in two, for all practical purposes.

The real benefit from these braces is that they carry the weight of the center of the roof out to the ends, where the top log chords can brace it. They create a pair of three-dimensional triangles leaning into each other. So there's no swaybacked roof, no bulging top plates, however thick, from roof weight pushing down and out. This can also be done with collar ties, which are rafter chords above head height. But these, being high, still allow outward spring of the rafters.

When building the Kruger house in Missouri, which had no collar ties, two carpenter friends working with me doubted this strongly — more so when I announced that we were going to drop king posts from the peak, becoming studs for a dividing upstairs wall, to support the 20-foot span of the ceiling joists.

Student-built cabin in Bethel, Missouri. Extended plates will carry pairs of trim rafters for the eaves.

My skeptical friends chained the log plates together in the middle to keep them from bulging, then went ahead with the bracing, roof decking, king posts, and upstairs subflooring (which was also the downstairs ceiling). They reminded me often of the weight we were adding, and of my folly in general. Finally we loosened the chain to the accompaniment of only minute creakings. Everything stayed true.

I will point out that these braces must be spiked into place very securely, and each board of roof decking or slatting must be nailed to them as well as to the rafters. In a large house, they can carry several tons, all to the corners, where the logs themselves would have to stretch and the gables collapse for the roof to sag. In the Kruger house, I used steel pins at the corners to better take the stress.

True, the pioneers didn't do any of this. (However, some of the better log builders did.) If you do, your roof won't sag, lean, or domino over in your average typhoon. You may also deck the roof with plywood instead of board slatting to act as a continuous angle brace.

Either king posts or queen posts are important to keep the ceiling joists from sagging if the span is long, say over 18 feet. No matter how massive, these joists will tremble underfoot unless braced in the center. The alternative is posts downstairs, right in the middle of things, or a summer beam.

This beam performs exactly the same function as the sleeper under your floor joists, but to be effective it should be braced with a post at least at midpoint. These beams were common in New England clapboard houses, framed as they were in heavy hewn oak. There's a good feel about a summer beam overhead, holding up the joists, but it reduces headroom, too. They were rarely used in mountain cabins. We used a 32-foot one to brace the joists for the 26-foot span in the log house at Wintergreen.

Bird-Stop Boards and Soffits

If the rafters form eaves, you'll have a gap between the top plate and the rafters. To seal above the top log plate between the rafters, cut and fit boards — known by the delightful name of bird-stop boards. Each board, nailed flush with the log, extends to the slope of the roof. I use one inside and out, with insulation packed between. By nailing these boards against the sides of the plate, any unevenness on top caused by variations in rafter mortises (to compensate for top plate unevenness) is covered. If you've terminated the rafters at the plate, this isn't necessary.

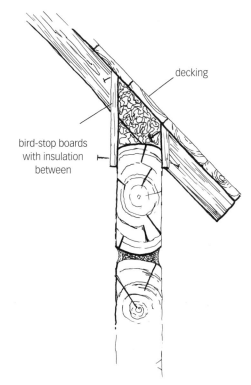

bird-stop boards with insulation between

decking

Bird-stop boards seal the spaces between rafters where they pass over the plate logs. In conventional houses, this area is boxed in with a soffit and rake.

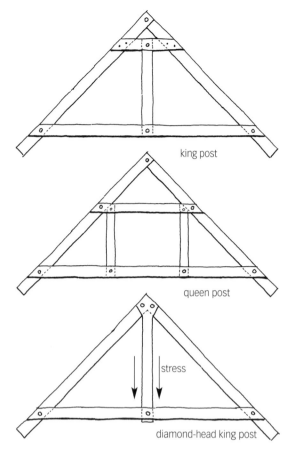

king post

queen post

stress

diamond-head king post

Either a king post or queen posts were necessary to support a long ceiling joist so it would not sag under upstairs floor weight. The king posts divide the loft down the center, which is good if the house is deep enough for two rooms. Narrow houses sometimes used the queen posts, which open a passage down the center of the loft.

If you prefer, rafter ends may be cut plumb and soffits extended back to the top log face, as in conventional roof construction. A vertical rake board is nailed over the rafter ends, usually extending an inch below the soffit board. With hewn-log top plates, this soffit must be scribed to fit the uneven log face.

Roof Slatting/Decking

Let's get things under the roof. If you're a purist, you'll slat your rafters with split boards, poles, or rough-sawn boards with the bark edges left on. For evenness, the settlers used sawn lumber here very early, because you get a much better fit with shakes on a flat surface. I know of many old houses slatted with edge lumber sawed with a pit or whipsaw, before circle sawmills came. These old saws were powered by muscle, water, or steam, and utilized a long blade that went up and down through a horizontal log.

I nail slats, which are 1×6 or more, about four inches apart. This lets air circulate around the shakes to help keep them dry and holds off rot. With a metal roof, it also keeps moisture from condensing underneath. Screen the spaces between slats at the eaves to keep out unwanted creatures. We use galvanized chinking wire mesh here.

Of course, purism notwithstanding, a slatted roof with wood shingles *will* let in cold air, hot air, wasps, lizards, spiders, and, if you're not the best shingler, some water. The alternative is solid decking, with a strip of tar paper laid under each course of shakes. Solid decking gives you a mite of insulation too, but not enough to count on. It does keep the great outdoors out there where it belongs, however. Now we often deck solidly, then slat above to let air circulate between the decking and the permanent roofing.

If you plan to spend more than fair-weather holidays in your house (and you will, no matter what you plan now), deck solidly, insulate, and tar-paper under the shakes. If you want to see the undersides of those shakes between the slats, slat just your porch roof, where it doesn't matter as much.

Use either plywood or wide boards when decking. Plywood is heavy to handle by yourself, but it goes up fast. Wide boards, say 1×12s, are easier to handle, but they cost more. We slat where we can or use ⅝-inch plywood, if the owners insist.

I like 1×12 boards, which also go up fast. You can find large lumberyards that stock a utility grade in long lengths that's cheap and usable, if you pick and

Craftsman Eric Bolton installs porch eave soffit boards.

choose a bit. I've been able to throw out a lot of it sometimes and still be ahead of the price game. Or get your local saw miller to cut you some decking. If put up green, it'll shrink, but cracks here don't matter.

Tongue-in-groove lumber is ideal, but expensive. The two-inch-thick variety is available most places, and will span up to six feet easily at a 45-degree pitch. Most people who buy this think it'll be extra insulation, but again, it doesn't help much.

Trim the ends of the decking with an eave rafter of the same size as the main rafters, carried on the extended plate or lookout.

If you can, build with the top log plates extended to support this eave rafter at the bottom. Traditionally, this bit was avoided because older houses didn't have much eave. When mud-and-stick (catted) chimneys were used, the plates and ridge were extended so the roof could shelter the chimney.

Insulation

After you've decked with your chosen material, you can staple down a layer of sheet plastic or tar paper to seal the roof temporarily against rain while you work inside. But take this off when you apply the permanent roof if you insulate — which you should.

Insulation should be of a type that traps lots of tiny air pockets, as with Styrofoam, mineral wool, or fiberglass. You can insulate between the rafters and finish inside with some kind of wall if you don't want to look at exposed rafters. Or you can put a nailing strip along each rafter, insulate between with a layer of Styrofoam, and cover with a recessed wall to leave part of the beam showing. Either way, you don't want a solid barrier, such as plastic or tar paper, outside the insulation because moisture will condense. The insulation should be at the same temperature and humidity as the outside air.

You can nail upright beams, say 2×6s, onto the decking outside for a built-up or double roof. Insulate between beams, again with something like Styrofoam, which does an acceptable job in relatively thin applications. Next nail slats. And then nail the shakes or metal roof to the slats. I like the Styrofoam outside because it's waterproof. The best grade is the dense blue stuff, which is more expensive but better insula-

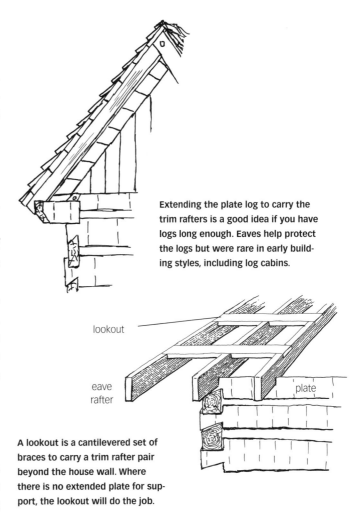

Extending the plate log to carry the trim rafters is a good idea if you have logs long enough. Eaves help protect the logs but were rare in early building styles, including log cabins.

lookout

eave rafter

plate

A lookout is a cantilevered set of braces to carry a trim rafter pair beyond the house wall. Where there is no extended plate for support, the lookout will do the job.

tion and lasts longer. Fiber insulation that's wet is no insulation at all. Also the foam insulates better per inch thickness, and you don't get particles of it in your lungs, which stay forever.

Again, the early settlers didn't insulate. If ceiling boards, with maybe handmade rugs or bearskins over them upstairs, held heat down in the main rooms, the kids slept among the elements. Later, everything from old newspapers to burlap sacks and cheap wallpaper was applied to inner walls and roofs, but its purpose was mainly to stop wind, and it did not impede the flow of heat through solids.

Years ago, I used a couple of layers of Celotex, a wood-fiber sheeting, as insulation on top of solid decking. It was fair, but insufficient. Remember, you'll lose most of your heat at the top, so trap it there. An insulated ceiling will keep things warm downstairs if you don't use the loft for anything but storage. But if you live up there, insulate the roof, one way or another, and let that heat go up through the ceiling. We use at least R-19 overhead.

I like to see exposed beams everywhere, so I prefer a solid deck, with insulation on top and shakes nailed to slats (built-up roof). Don't nail down roofing through soft insulation. Shakes move around in wind and temperature changes, and can work loose.

Roofing Prices

On a price scale, from lowest up, roofing materials run like this:

- asphalt shingles
- barn tin (two-foot-wide sheets of corrugated or five-groove nailed with washered nails)
- standing seam (hammered) galvanized metal, two-foot-wide runs
- split wood shakes
- standing-seam terne tin (which must be painted right away), 17-inch runs
- standing seam copper (whose color mellows to a dark neutral brown and eventually to green), two-foot runs
- slate

I refuse to roof a cabin with asphalt shingles. I have done slate, and I love to work with copper, but they are so very expensive. Any attempt to give a price on these roofs installed would be far off, because costs vary so much in given regions. We've begun doing our own roofs entirely because the labor is so high.

We do standing-seam galvanized for about $425 per square (100 square feet). Terne is about $500 a square and another $130+ a square to prime and paint it. Copper is $725+ a square depending on the current per-pound cost of copper, but it never needs painting. Cedar shakes done properly are about $450 a square.

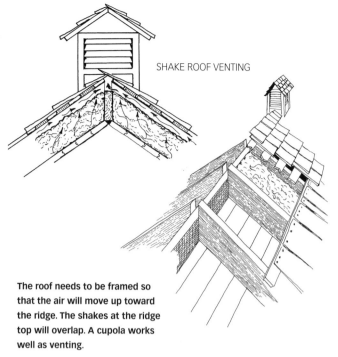

SHAKE ROOF VENTING

The roof needs to be framed so that the air will move up toward the ridge. The shakes at the ridge top will overlap. A cupola works well as venting.

Dormers, valleys, and multiple chimneys all increase roofing costs. So do very steep roof pitches and very high roofs. We do only the roofs of our own buildings and restorations, and when pressed for time or have a very complicated roof, we will subcontract to a good roofer.

In Virginia and the East, metal roofs have been around so long they've become a tradition. We do more of them than shakes now. Modern tools for installing them, though, are very expensive and old manual tools are very slow and labor intensive. I don't advise doing your own hammered metal roof unless you get experience elsewhere.

The sheets of metal with washered nails are seldom used on houses, but often on barns, workshops, and sheds. If galvanized, they, like standing seam, should

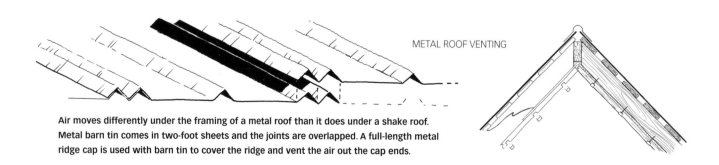

METAL ROOF VENTING

Air moves differently under the framing of a metal roof than it does under a shake roof. Metal barn tin comes in two-foot sheets and the joints are overlapped. A full-length metal ridge cap is used with barn tin to cover the ridge and vent the air out the cap ends.

be painted eventually, after the shine is gone. Eventually, you'll need to seal or caulk each nail head when the lead or plastic washer fails. Standing-seam metal leaves no nail heads to the weather.

Wood Shakes

Shingles are sawn or cut; shakes are split or riven. If you split your own, good luck, and tell us how it came out when you finally finish. I'm a pretty fast hand at riving shakes, but I cannot compete with the price of lumberyard shakes or shingles. With practice, a wiry individual can split several hundred shakes a day from prime timber, but *that* is precisely the biggest hurdle.

The settlers used cedar, oak, chestnut, or cypress if they were far enough south. That virgin growth was fine, but today you're hard-pressed for good, sizable, clear wood. It must have no knots, be straight, be two feet or more in diameter, and be close grained.

Say you do have a load of two-foot-long oak blocks. Red oak is the easiest of the oaks to split, but not the most durable. You quarter the blocks, using a sledgehammer and wedges. Then split these in two and you have eight very tall pie slices. Split off the point at the heart, and the sapwood and bark, leaving maybe eight inches or, if the block is big enough, maybe wider. Now use a froe, an L-shaped tool, to split the shakes along the radial split lines from the heart. That means each shake is a little thicker on the outside, but it's also the only way to do it. The old-timers dressed these with a drawknife on a shaving horse.

Drive the froe with a hardwood mallet, ¾ inch or ½ inch from the edge of the pie slice. Or split the block in the middle, over and over till you have all shakes. Wedge the block into something like the forks of a splitting or riving horse and pry off the shake with the froe handle. With practice you'll learn that prying one way runs the split in and the other runs it out. At first you'll be lucky if the split gets all the way to the bottom without running out. If this keeps happening and you're getting a narrow top and wide bottom to your block, turn the bottom up and work on that end for a while.

You can split shakes green or seasoned; there are craftspeople who insist that only one or the other will work. Just be sure they season before you nail them

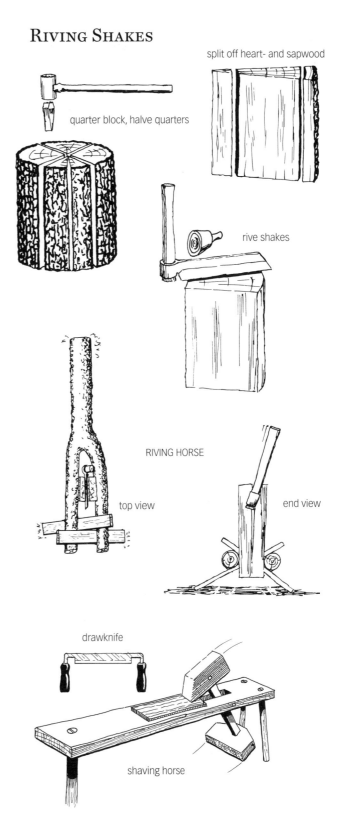

split off heart- and sapwood

quarter block, halve quarters

rive shakes

RIVING HORSE

top view

end view

drawknife

shaving horse

The split shake, or any other piece of wood to be shaved with the drawknife, is clamped in place with foot pressure. The woodworker sits on the shaving horse and pulls the drawknife toward him.

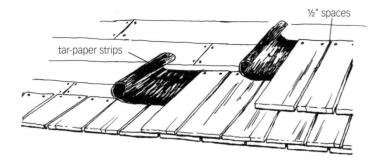

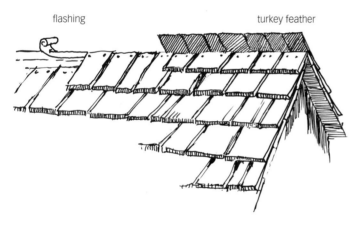

cap

Two peak treatments, both with metal flashing under. The "turkey feather" is traditional.

flashing   turkey feather

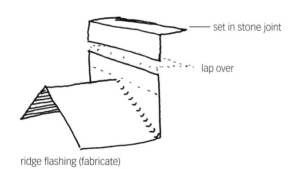

set in stone joint

lap over

ridge flashing (fabricate)

This peak flashing against the chimney should be fabricated so it can bend in more than one direction. This is a common source of roof leaks if not done properly.

½" spaces

tar-paper strips

Underlay shakes with 30-pound felt paper, so that any leaks will run out on top. Water trapped under the tar paper will rot decking. The first row of shakes is laid double.

up. Shrinking will split them right where the nails go in. There are astrological signs best for this activity, as there seem to be for just about everything else, so if you're into this sort of thing, lay them up at the increase of the moon, by all means.

Cedar, cypress, and pine are worked differently. Square the block first, then go right across the middle with the froe, one shake at a time. If the block is large, halve it first. Oak, with its strong radial split lines, will crack down the middle of the shake if split across the grain this way, from "shake" cracking.

A word about these woods — second-growth pine will rot quickly unless treated well with preservative, which won't last long itself in the rain and sun. Cedar, of the native red variety, is fine if you trim the white sapwood off. The western red cedar sold in lumber-yards is not as durable as the native red cedar. Cypress is fine, where available. Black walnut would do nicely, but let's not kid ourselves.

Cedar that can be riven is almost nonexistent today, except in protected groves, whose owners frown upon its removal. Even red oak is hard to find, so I've switched largely to lumber company offerings and sawn shingles. Even knotty cedar can be sawn nicely. And oak, if carefully done, can be sawn. I prefer the cedar, partly for the smell, which I never tire of. Cut shingles, sliced with a large mechanical knife, can be bought also, but are quite thin.

In estimating the number of shakes you'll need, fig-ure the square footage of the roof, including eaves, and add maybe 200 square feet for waste. Shingles and shakes are sold by the square or bundle, with a square being 100 square feet, five bundles to a square. If you split your own, figure a mountain of blocks, then mul-tiply that by 10. You still won't have enough.

Lay a double thickness of shakes for the first course at the bottom, to cover all the cracks. You may want to cut the first layers in half. Then, lay a strip of tar paper half the width of your shake length under the upper half of each course. You should, ideally, leave exposed only a third of each shake, but on a 45-degree roof you can cheat a little. Don't leave as much as half of the shake exposed, though. With tar paper under, using 24-inch shakes, you can leave 10 inches exposed, which still gives a 4-inch overlap on every other course. That's important, because each shake must cover the crack of half an inch or so left between the shakes of the course below it. This means that, with random-width shakes, you'll use about the same width up any given section of roof.

Clear? Well, you can't cover a layer of eight-inch-wide shakes, half inch spaced, with four-inch shakes unless you leave huge spaces between. You may use the narrow ones you'll inevitably find in your bundles to double up in covering extra-wide ones. And, of course, these narrow ones are necessary at the ends, to come out even. Leave a half inch or more overhang at the eaves to keep the decking dry.

Nail with two shingle nails or #6 galvanized nails per shake. Cover the nails with the next course and they won't rust anyway.

At the top, you'll find it necessary to cut at least one course of shakes, probably two, then finish with a layer on each side laid horizontally over roof flashing. Or go traditional, and leave the last layer of one side jutting up, away from the prevailing winds. I'm told that was called a turkey feather roof, and the worn shakes standing skyward on old cabins do look like bedraggled feathers. We do this on authentic projects, with flashing under; there is no problem with leaks.

Standing-Seam Metal Roof

More of our cabin roofs today are standing-seam metal than wood shakes. This roof, common for 150 years around Virginia, Maryland, and Pennsylvania, involves using "pans" of sheet metal that are bent and crimped together, along with the nailed-down clips that hold it all in place. No nail heads are exposed with standing-seam metal, although installing it is a labor-intensive operation. Barn tin, corrugated or 5-V, is the same metal, but is nailed with washered nails in the ridges to discourage leaks.

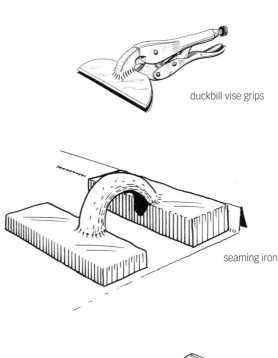

duckbill vise grips

seaming iron

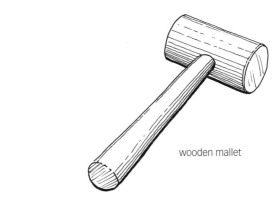

wooden mallet

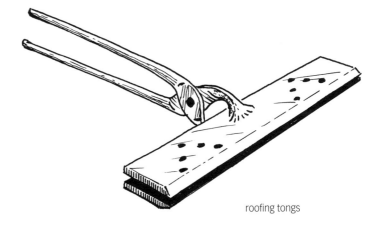

roofing tongs

BUILDING A STANDING-SEAM METAL ROOF

Standing-seam metal roofs have been used for so long in the eastern and central United States they have become historic. For many years, the metal was not available in this country, and was imported from England in short lengths. Now it is available in rolls of galvanized metal or copper, as well as a modern version of English tin-coated sheet iron called terne. Except for copper, the metal requires painting, however. When painted, a metal roof lasts longer than wood or asphalt shakes.

The metal is installed over spaced or solid roof decking, with red rosin paper or asphalt felt paper between. The method of installation leaves no nail heads as points for potential leaks, and where properly applied, no soldering or caulking is necessary.

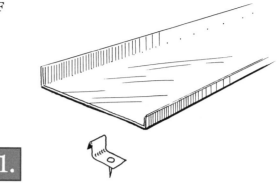

1.

The procedure involves bending up the edges of the lengths of metal at two different heights of the bend, forming "pans."

2.

The higher of these is placed next to the lower of the adjacent one. The lower has metal "clips" nailed next to it and the clips are bent over the raised edge. (Clips are usually made on the job site by cutting narrow strips of leftover metal into pieces about four inches long.)

3. The higher edge is bent over and crimped over the lower, incorporating the clip and covering the nail heads. A second bend and crimp locks the pan in place.

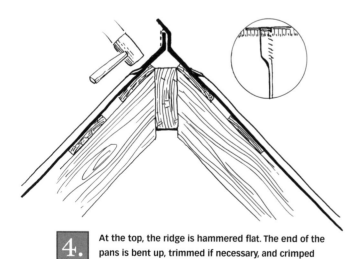

4. At the top, the ridge is hammered flat. The end of the pans is bent up, trimmed if necessary, and crimped over the pans from the other side of the roof.

5. The bottom edges are bent and crimped over specially formed drip-edge (usually bought in 10-foot strips when purchasing the rolls of metal) nailed in place. Like all roofs, a metal roof best protects the house in conjunction with a good guttering system. Both the roof and the gutter must be of compatible metal, or they will corrode.

Flashing

Use flashing in valleys, around dormers and wings, around chimneys, and on all ridges. The aluminum kind is cheap and pliable, and less trouble than old off-set printing plates or flattened cans. We use copper flashing today, with copper nails, because it lasts longer. Copper tones down to a neutral brown; none of it shows. Always use the same kind of metal in nails and flashing; dissimilar metals corrode each other.

Chimney Space and Flashing

Your chimney will go through the roof somewhere, regardless of the positioning of the fireplace in your house. You may have thought to leave a section of decking out for the chimney to come up through. I never do; I saw out the section later for a better roof-to-chimney fit. Anyway, don't roof this space, usually a three-foot-wide cut in the eave dimension.

Guttering

Guttering your cabin is a necessity. It keeps runoff from soaking the ground under the foundation and helps keep water from getting into a basement. Gutters aren't pretty, but they're useful. We use half-round, usually copper, gutters with downspouts piped away from the house.

Remember not to let dissimilar metals touch; a metal roof and copper gutter straps will corrode each other. This applies to all places where you might use metals — wall flashing, porches, snowbirds, chimney flashing, conjoined roofs, etc.

Also, if you gutter, you'll want to use snowbirds to hold the snow till it melts, so it won't tear the gutters off. Again, use similar metals — bronze for a copper roof, zinc alloy for a galvanized roof.

By all means, install a permanent screen over the gutters to keep out leaves. Decaying leaf acid will corrode any metal.

If you do not have access to commercial guttering, or if you just like the look of wood, you many build gutters of wood. Cypress is expensive, but long-lasting. Pressure-treated wood will also serve. A vee-trough is

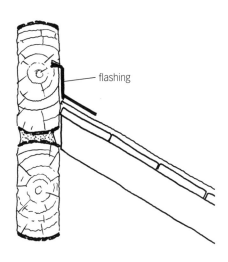

Flashing is imperative when a roof for a porch or lower addition abuts a log wall. Cut a groove into the log one inch deep and insert the flashing the full length of the roof. Use copper if the roof is made of shakes or copper. Use galvanized metal or aluminum if the roof is made of that metal.

simple, supported on brackets of wood or metal. However, downspouts are harder with wooden gutters.

The Gables — Outside and Inside

Close the gable ends in conventional stud-wall fashion, laying a 2×4 or 2×6 sill on the top log if you want it more even. With round or unevenly hewn rafters, you can spike another 2×4 or 2×6 to the undersides of these to nail the gable wall to. Frame windows conventionally, remembering to leave space at the fireplace end for the chimney.

Clapboard, board and batten, and shakes were commonly used for the exterior of the gables. These walls should be insulated. And you will want to finish them with interior wall covering. Here, I favor paneling grade of 1×12 lumber used vertically. You'll add horizontal 2×4s, called fire-stops, to nail to. It's a simple operation to run this wall material through a spindle shaper or dado blade to shiplap it. Or you can purchase tongue-in-groove lumber, which is more expensive. Either will keep cracks from opening up if there is shrinkage.

With your roof complete, you're just about halfway through with the house. At this point, the pioneers lacked only a fireplace, chinking, door, and maybe a floor, but things are more complicated today. You still have windows, stairs, wiring, plumbing, cabinets, boxing in the lean-to, and just lots more.

But the roof is on, so you can get to other things. Start moving things in out of the weather. And, of course, you can relax your frenzied pace a bit now that the lid is on, and work along so as to be ready for chinking about the time the logs season and settle.

Our Missouri cabin gets its sawn cedar shingle gable covering. The roof is temporarily covered with tarpaper.

ROOF CHECKLIST

☐ Don't work on a roof by yourself.

☐ Space rafters 24 inches o.c. or less if using 2×10s (for insulation to meet code). If not insulating that heavily, you may want to use 2×6s or 2×8s, 16 inches o.c.

☐ Make sure top plates are level.

☐ When measuring rafters for the plate notch, mark the apex of the notch.

☐ Lay out the rafter pairs, or trusses.

☐ Level vertically, brace temporarily, and peg to the plates.

☐ Place the roof bracing as described in the text.

☐ Cut and fit bird-stop boards to seal the gap between the top plate and rafters.

☐ Choose a method of roof decking.

☐ Insulate with Styrofoam, mineral wool, or fiberglass.

☐ Choose your roofing material.

☐ Add flashing in valleys, around dormers and chimneys, and on ridges.

☐ Install guttering and downspouts all around.

A Gallery of
Hewn-Log
Houses

The Pioneer Cabin

The restored John Oliver cabin is located in a 6,800-acre section of the Great Smoky Mountain National Park known as Cades Cove, near the cities of Townsend and Gatlinburg, Tennessee. Oliver and his wife, Lucretia, were among the first European settlers in Cades Cove, in 1818. The cabin was built for their son in the 1850s.

PREVIOUS PAGE: A handmade heart-pine door
with forged hinges and latch.

Restorations

The Snyder cabin was built from oak and chestnut logs. It was moved and reassembled on the shore of a pond and is used as a retreat.

The McRaven house near Free Union, Virginia, is actually three recycled houses joined together — a log cabin and two unusual post-and-beam houses. Stone completely covers the kitchen end, and stone-ended stucco covers the other end, which is the living room. The family room and master bedroom are in the log section.

Restorations

This house at the base of Virginia's Blue Ridge Mountains was built to be as maintenance-free as possible; its materials include vintage logs, stone, cedar siding, bronze screening, and rough-sawn oak. The kitchen/dining room is housed in the stone section, and the living room and bedroom are in the log section.

During the complete on-site restoration of the Beadle house in 2002, the owners added a frame addition for the kitchen and dining room, which has a wonderful view of the lake and rolling backyard. They also added a small two-story log cabin for the new master bedroom suite and a guest room.

The weekend house at Culloden Farm was moved in 1979 to this hilltop above the Kentucky River. The chimneys are of the local blue limestone.

This cabin in Wintergreen, Virginia, combines old chestnut logs with a new frame wing and full basement. The result is a cabin that successfully integrates traditional building materials with contemporary design and function (complete with a hot tub on the back deck). The owners commemorated the restoration with a plaque thanking the McRaven crew.

Before & After

BEFORE

AFTER

The 1769 Sam Black tavern was moved to a new site in 2002 after a property transfer. Documented in Thomas Jefferson's diary, the tavern is now on the National Register of Historic Places.

BEFORE

AFTER

This chestnut-log house was built by Captain Beadle around 1788. The new owners had the vision to see past the complete disrepair and the overgrown site (which was cleared during construction). The house is now on the National Register of Historic Places.

BEFORE

AFTER

The previous edition of this book inspired Philip and Teresa Babcock to take on the challenge of a complete restoration. They later wrote me a letter about it, and told me what a satisfying accomplishment the adventure was.

BEFORE

AFTER

The abandoned Page Meadows house near Free Union, Virginia, dates to 1770 but was hidden inside an 1859 farmhouse. The log house had been engulfed over the years with timber frame additions, clapboarded on the exterior and plastered over on the interior. In fact, the most recent inhabitants never realized they were actually living in a log house.

The Perfect Site

This cabin is a mountaintop retreat near McDowell, Virginia. The beautiful mountain views and rolling hills are a trade-off for the exposed, windy site.

Choosing a site for your house, whether on the banks of a pond or tucked into the woods, must take into account your needs, as well as how the house will be used. Hewn-log houses look best when they appear to be a natural part of the landscape.

When selecting a site, keep in mind the view you'll have from the front porch (often the most lived-in part of a hewn-log house) and the exposure to sunlight the house will receive.

The restored Sam Black tavern (see page 124) shown at its new site, on a farm in Virginia. The split-rail fence is an authentic finishing touch.

The Possum Creek cabin was half a dogtrot restored in the traditional style. Nestled in the woods, the cabin was built to be a weaving studio and guest quarters near the main house.

Porches for Outdoor Living

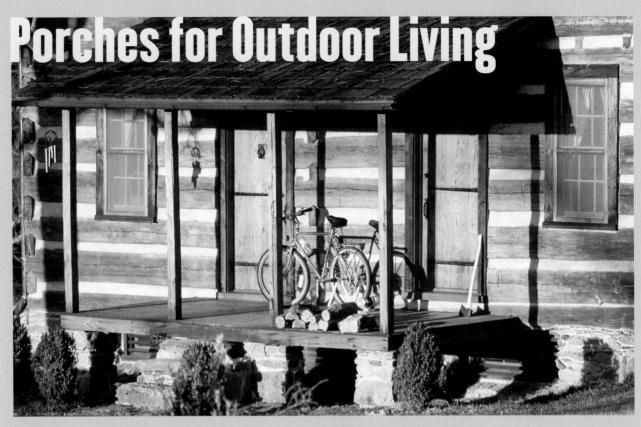

The two chestnut front doors on the Page Meadows house are customary for the age of this cabin. The simple porch is supported by piers of mortared stone.

Comfortable rocking chairs and flowers from the garden create a welcoming scene for those looking for a place to rest.

This screened porch is so well protected, its owners have essentially created an additional living room.

The guests in this cabin have a stunning view of the Smoky Mountains from the rocking chairs on their porch.

The Snyder porch bench invites the gardener to take a break.

The Bonner cabin porches were built to open onto the spectacular view of the Hardware River Valley and the Blue Ridge beyond. The double porch was often called a "sleeping" porch. The porch on the second floor is accessed from the master bedroom.

Interior Options

Light is an important factor in a log house. This master bedroom has gable-end windows as large as possible, dormer windows and a specially built shed dormer facing the pond.

The Wintergreen cabin dining room was in the log addition to the main log pen. It has been the site of many family get-togethers for the owners since 1985. The white tile and cabinetry of the kitchen (RIGHT) is a pleasing contrast to the chestnut logs and beams.

The use of open ceiling beam work in this more formal living room creates an airiness that counter-balances the weight of the log walls. As a result, the room feels substantial but not dark.

The owners of this log house have chosen to paint the logs and chinking white, to give the room a lighter feel.

A rustic style is easily accomplished in a hewn-log house. This bathroom is complete with chamois curtains and a tree trunk for a towel holder.

This nook offers a comfortable, well-lit place to read.

The stairwell of the McRaven house combines complementary textures of log, stone, and planed and rough timbers. The treads are of three-inch-thick heart pine recycled from an old Lynchburg, Virginia, cotton mill.

The huge logs used in building this house feel substantial and rightly dominate the aesthetics of the interior.

Under Construction

The author adzing a replacement log for a restored cabin of white pine logs near Earlysville, Virginia.

Willie Lehmann cuts a half-lap joint for a log repair, keeping the notched part of the log in place.

The chimney under construction on a relocated rustic cabin near Branson, Missouri.

CHAPTER TEN

Windows and Doors

ALONG WITH THE TREASURED PIECES of
china, wrapped perhaps in a feather bed to withstand
the jolting of the wagon, was often a pane or two of
glass for the settlers' house. Handblown, wavy, thick,
and so fragile, it traveled west to grace the log house,
and was often removed if the family moved on.

Early Windows

Many cabins were built with no glass. The window
openings were shuttered with split or sawn boards,
leather-hinged, against the cold. I know of one log
house near Gainesville, Missouri, built around 1900
that has only one window with no glass; its shutters
hung on forged-iron hinges. The Murphys' house,
though only 40 years old, had just one small, four-
paned window.

You see, wood was plentiful. Unless it was chopped
continually and used, the forest would cover a homestead in a few seasons. Iron and, glass were rare and
prized. That is why early houses had small windows,
usually 4-over-4 double-hung or 6-over-6 with 8×10
panes. Windows were generally hand built until the
1840s, when factory-made windows and doors appeared. I know of factory double-hungs in an 1838
house on the James River in Virginia.

Even where several shuttered windows were cut,
with no glass, they were always sources of drafts and
insects, and were troublesome to build. So windows
were few. But today a log house need not be dark to be
free of drafts. But remember, the heat escapes through
glass easily, and that too many windows leave short
lengths of logs, which weaken the log structure.

To retain the traditional log cabin look, windowpanes should be small
and of the same style throughout the house.

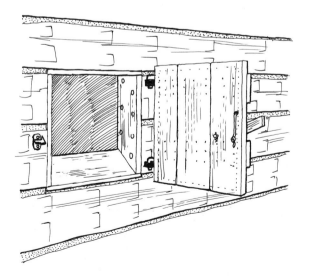

Shutters like these were often the only light source in early log cabins.

A bay window in a log retreat. Such innovations make log interiors light and cheerful, but there is a great deal of heat loss through this much glass.

Hand-built window on a round-log cabin. These units can have a lot of character, but require a great deal of time and labor to construct.

Choosing Your Windows

Windows (and doors) are probably the single most expensive building materials you will buy. They involve an aesthetic decision, too. Careful planning and purchase of windows could well be the best investment you make (other than good logs, of course). Whereas old windows seem to have style and charm, today's manufacturers can build both beautiful and energy-efficient windows. Be aware that there are many cheap and/or ugly ones out in the market. Nothing blows the look of a house like inappropriate windows.

When making your decision, you have several choices. You can build your own, which is a difficult, time-consuming, exacting chore best reserved for special windows such as a stained-glass fixed window. Or buy barn sashes and install in the window frame you built; then add locks, screens, and storms. You can also reuse salvaged windows. However, repairs often make this inefficient and expensive. Finally, you can buy new, commercial units. These come complete with screens, insulating panels or storms, pop-out frames for easy cleaning, locks, opening mechanisms, and so on. Your budget determines your choice, but this is no place to skimp. Delay a finishing touch or conserve on something else so you can put in the best windows.

Log house builders have always favored small windows, either to save glass or for the more valid reason that large cuts in a log wall weaken it. So the double-hung sliding sash has limitations. Also, you ought to keep in mind that if there is enough space for a wood sash to slide, there is enough space for drafts to enter. Modern double-hungs usually run on aluminum or plastic runners and are draft-free.

The Barn Sash

The cheapest window you'll find for your house is called a barn sash. It's just glass and wood, one piece, with as many panes as you like. I prefer small panes, which are easier and cheaper to replace, and they look more traditional in a log house. I swing them on hinges, against felt-lined stops cut ¾ inch square and nailed to the window jamb.

LEFT **These shop-built French doors have dual glass sealed panels in both the doors and the transom lights over the doors. The heavy muntins are necessary to handle the extra weight.** RIGHT **The single barn sash is perhaps the simplest log cabin window. Close it against weather stripping with a cupboard latch.**

It's simpler to hinge windows so they swing inward, leaving room for screens outside. If you need the space, swing them outward, but you'll need some method for controlling them through the inside screen. I have yet to fabricate a satisfactory low-cost device, but you can easily buy crank-out controls from your local hardware store.

Everyone I know who has a log house with hinged windows wants hand-forged hinges. I usually install barn strap or T-strap hinges temporarily, and sometimes I find time to replace them, one by one, as I can get to my forge. At this writing, I still have not completed this job in our own house.

Latches, too, can be simple and inexpensive. I use cupboard latches on windows, of the type that lift and drop into a notch, but these must be good quality to wear well. A sliding thumb latch will also work. Even simpler is the screen door–type hook and eye.

Commercial Window Units

You can, of course, avoid just about all the details of window and screen installation by buying the completely assembled units. They come with screens, hinges, and latches — all boxed into jamb framing that is simply inserted into a hole in the log wall, shimmed

level, and nailed into place. Casement windows come with devices for cranking the glass open.

Suppliers have unlimited design offerings, but standard sizes vary, so make sure you can get the windows you want. Any supplier will custom-build for a price. You will certainly want wooden windows, not metal or vinyl or clad windows.

Again, I will warn you that these windows can be the single most expensive item of your cabin. But having draft-free, attractive windows adds to the pleasure of living in your house. And by this time, you may be at a stage when your time to do other things is worth more than the extra expense.

Window size is a matter for your own taste, within certain bounds. A huge single-paned picture window is out of place in a log cabin. A double window works better, and often will solve another problem: Some of the decayed log notches or sections in that wall may have to be recut. This means shortening the logs and enlarging the window space for the doubles.

Most people do prefer the double-hung windows, which come in several grades. Wooden six-over-six pre-hung units with screen and single-thickness glass were the norm for 150 years. Now, single-thickness glass has to be special-ordered, and costs more than dual glass. Better name brands — like Marvin, Pella,

This cabin created from two separate pens has the second front door converted to a window. Additional windows in the knee wall let in light from the west.

Kolbe — are much higher in price, but the workmanship and materials are better.

Today, most building codes require double-thickness windows. Check your area for confirmation. When you make the switch from recycled single-thickness glass to new double, the price goes up dramatically. Your local weather will help you decide if the cost and the look of the better brands are necessary. Double glass cuts heat loss, although it's still high. The thing you'll notice first is that the reasonably priced double-glass (insulated glass) windows will have no muntins (the dividers between panes). They come with some plastic or wood snap-ins that don't look remotely real. Other choices include real divided lights, which are muntins heavy enough to take the dual-glass sandwich. These are the most expensive windows.

You can also get winter weather protection from recycled single-pane-glass, divided-light windows if you make careful choices. Limit the number and size of windows on the north face of your house and add more to the south face, to take advantage of solar gain. Plan for detachable storm windows over divided-light sashes. Some companies make snap-in or fitted panels (smaller storms that don't cover the whole window assembly) to create double-glass protection.

The better windows have features like pop-out sashes for cleaning and more precise milling of better-quality wood. The cheaper ones usually do all right if you heat with plenty of wood and don't air-condition. We have always used dual glass with electric heat or air-conditioning to cut heat loss or gain. The two or three times the cost eventually saves enough on energy to make sense.

Ironically, about the time milled windows became common in the remote areas, the style had changed from the earlier classic small-paned ones to the large, single-paned sashes of the late Victorian period. So, many cabins were given these windows later, and many built in the later 1800s and early 1900s had single-pane windows.

I see single glass by the 1870s. I examine the workmanship carefully because these sometimes are replacements. By this time, a popular style was a two-over-two, which replaced earlier multiple panes.

Keep the same style, if not size, windows in the main house and addition. If your house is large enough for dormers, use the same style sashes there too.

Dormers

Dormer windows do special things to the outside and the inside of a log house. Outside, they look right. I like a dormer to be tall and slim, and show off a window with style. The dormers in our Virginia house have handmade barn sashes latched with simple catches for the main windows. In the fixed arch at the peak, the curved- and cross-arched mullion semicircle looks especially striking lit at night.

Inside, dormers allow light to enter your house at interesting angles. They also keep the interior wall finishing from being flat and boring — to say the least. Waking in a room served by dormers is a very pleasurable experience.

Dormers allow you to walk up to the outside wall in spite of a low roof slope, and give at least the feeling of more space. They are lots of work to build, however, and you should be a pretty advanced carpenter to attempt building them. I often compare constructing a dormer to building a miniature house — that's how much detail is involved.

The Kruger house has its dormers built between four-foot centered heavy rafters, with 24×37 sashes. Anything larger would be disproportionate in the 28-foot roof length. Here, dormers were almost essential, with the narrow upstairs divided into two long bedrooms, walled off from an open area at the head of the stairs. Without the dormers, only one gable window in each room would have been possible.

Sometimes early builders used small windows in the kneewall to let in additional light. We do this often in one-and-a-half-story cabins, but must use tempered glass here.

The Wintergreen house has four dormers in the open-beam roof, which give light to key areas, such as kitchen, entryway, and master bedroom. They are dual glass (wooden sashes with factory storms).

Our current Virginia house is log with a timber-frame wing on one end and a stone wing on the other. We have two stories plus a loft, lighted with seven dormer windows and three more in a log addition. These are arched up into the peak, and were built of remilled heart pine by woodworker Wilson McIvor. The sashes swing on hinges, but the arched sections, with curved, intersecting muntins, are fixed.

Dormers look good from the exterior of a cabin, and are a great way to bring light into a room. They are, however, not easy or cheap to build; framing and installation are complicated. Dormers also pose challenges for both the roofing and the flashing.

This four-window group combined with the French doors lightens the dining room of the Wintergreen house. Interior trim here is mitered and beaded of recycled "naily" pine.

When we shop-build windows, we find it's cheaper to send them down to the glass shop to have the single or dual glass installed. Sometimes we save wavy, blown glass and have an occasional pane put in for the antique look.

It's virtually impossible to varnish or oil window wood outside to keep the natural color. We oil the inner surfaces of our recycled old wood but paint it on the outside. Factory windows are soft ponderosa or other pine, and should be painted, inside and out.

A lot of kit log designs use big picture windows, arched windows, oval, or other free-form windows. This is more of the philosophy of a modern-shaped house adapted to logs, and I don't think it looks right with hewn logs. Some very attractive window treatment can be done in additions and non-log sections of the house, however. Use your imagination: Every builder has a perfect right to put up something weird.

Exterior Doors

The hand-built front door with my own design and craftsmanship for the hinges and Norfolk or Suffolk latch has become the hallmark of a McRaven building

or restoration. It is my statement to the owner that this house is unique. Often we incorporate a shape into the ironwork that has special meaning for the owner, such as a specific leaf shape, or heart or flower or spade. In one case, we put in a rattlesnake; in another, we added a banjo neck.

I always build the doors myself, sometimes with glass in the front and back doors for more light. I build them because I want them to be more authentic, and because milled doors seldom look right. Very old cabins had doors that were not angle-braced in the familiar Z-pattern, but had rows of many square nails through the cross boards to keep them from sagging. In time they sagged anyway, as generations of slamming worked the nails loose.

So the Z-pattern became common, particularly in the South and West, as a simple and attractive solution. If you think this has a barn-door look, though, and don't want the nailed version, you may want to try your hand at a paneled door.

Our "standard" cabin door today was copied from a late-1700s house in the Blue Ridge Mountains. It is made from t.i.g. vertical boards with dovetail grooves about five inches wide for cross battens. Two or three

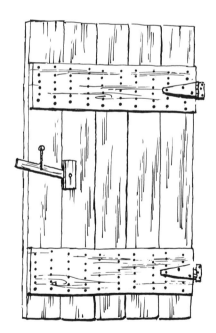  

Early cabins had vertical board doors with cross battens. Many nails were used to keep the doors from sagging. Later cabins used doors with the Z-pattern brace. The paneled door was used on cabins after factory millwork became available in the mid-1800s.

The short basement door was hand built, but the commercial hinges and latch fit the design well.

Hand-forged hinge showing pin.

of these are bevel-cut to fit the dovetail cutouts, and driven in from the side. Then forged nails are used in an X-pattern in each board to secure the battens. The door can't sag, and is more craftsmanlike than the Z-pattern.

I use heavy glass in outside doors, finding safety glass little more expensive and much safer than double-strength. Also, building codes require it. Using the router, I rabbet the window frame, caulk, lay in the glass, then set a thin wood strip against it with small screws. You can also use two strips with the glass between. Always build the door rigid in some way to keep stress off the glass.

I close the doors against felt-lined strips, as I do the casement windows. At the sill, I use a flexible weather stripping, which makes a stepped-up threshold unnecessary, although certainly an option.

A stable wood, such as white pine, is good for doors; yellow pine tends to warp. White oak is good, but heavy.

My favorite doors were for the Page Meadows house in Virginia, of 1⅝-inch chestnut, dovetail cross-battened and hung with forged hinges. These doors were grooved and splined. We hand-planed for texture.

Outside doors must be sealed to keep out moisture. Oil the inside, too, to avoid cupping and warping. I like tung oil, reapplied every four years or so.

The larger and later houses used commercially milled doors, inside and out, available generally for the past 150 years. They were usually pretty thin, and have never seemed to fit the solid look or concept of the overall log house.

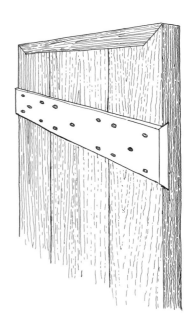

The cabin door we build most frequently is this one of vertical boards with the cross battens let in with a dovetail router bit. The battens are cut at an angle and driven tightly from the side to hold the door rigid. We always use it with forged strap hinges offset so the door swings in.

HAND-FORGED HARDWARE

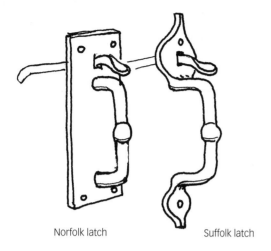

Norfolk latch   Suffolk latch

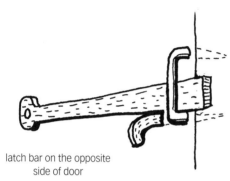

latch bar on the opposite
side of door

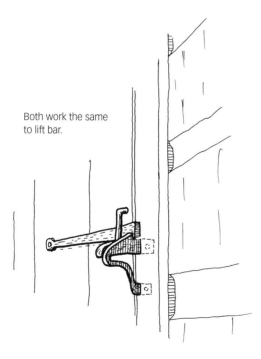

Both work the same
to lift bar.

Today, you can find every kind of milled door, some of which are massive and handsome. So, unfortunately, are their prices. If you buy new milled doors, do so beforehand so that you can plan specific doors for a specific place. The temptation to save work and time with low ceilings may become strong as you labor, which means shorter doors. But building codes forbid using many charming old doors that are too short (under six feet eight inches). Few modern commercial doors are constructed in a way that they can be cut down without severe weakening.

Hardware

Again, I like wrought-iron strap hinges for house doors, and these can also help brace the wood if they're long enough. Where a blacksmith was near, the old houses often had iron hinges, usually simple loop-and-pin straps. More ornate scrollwork was rare, unless the local smith specialized in such things.

I usually mount the hinges on the outside for appearance, and always have the cross battens inside out of the weather. This means offsetting the hinge so the pintle is inside for the door to swing inward. It's a simple bend when forging the strap. If the hinge is mounted onto the cross batten, inside, it's all too cluttered-looking. And the outside of the door is just this flat expanse of vertical boards.

For really heavy doors, I tried using three hinges, but it's really hard to line up the third one exactly. Now I use two, and forge them from metal at least ¼ inch thick, sometimes as wide as five inches.

A wooden hinge is a work of fine craftsmanship. Hickory or ash was usually used for toughness. It was greased with lard or bear fat, and served long and well. We reused a pair of these from a late-1700s log house in the children's Virginia Discovery Museum cabin we restored for a permanent exhibit in Charlottesville.

The wooden latch and string was almost universal until iron became common. Today I use a metal version of the old wooden latch, in either a Suffolk or a Norfolk thumb-latch pattern.

You'll find many cabins with the patent cast-iron door locks on the outside of the wood, sometimes with white ceramic knobs. (These knobs were sometimes used for nest eggs in the henhouse.) We have these

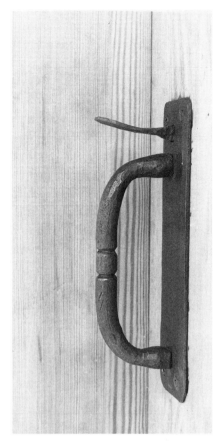

Handmade chestnut door with a Norfolk latch, forged strap hinges, and deadbolt. Exterior doors are sealed both inside and out to guard against warping.

cast-iron locks in our house and they are quite serviceable, except on heavy doors, where they tend to crack when slammed. One at a time, I'm replacing these with forged Norfolk latches.

A front door can be a really beautiful personal statement. Strong designs in carved wood or wrought iron as part of your door will set the style of craftsmanship that is inside.

Interior Doors

Interior doors were, of course, unheard of in one-room cabins. When used, I make these, too, often of t.i.g. of only ⅞-inch to one-inch thickness. I usually increase the thickness of the outside doors to at least 1½ inches. Here, recycled or salvaged matching doors can look very good. Building inspectors are often less stringent about interior door sizes, so interior doors become something you can play with.

Trim

Window and door trim in a log house often requires extra work. You have to provide a flat, even space to fit the trim. This sometimes means chiseling along irregular, extra-thick wall logs to let the trim in. Make this part of your window installation so that trim board fits exactly into the log.

Build door- and window jambs wide enough to match log thickness. You'll likely have to order prehung windows with wider jambs to match your thickness, because these will be for 2×4 stud walls.

Trim windows and doors with about four-inch-wide boards. Any wood is all right if you plan to paint it a tasteful color. Barnwood, grayed oak, or cedar boards are good if left natural. "Naily pine," or recycled, remilled wood with blackened nail holes, fits well into a log house interior. No clear finish lasts well outside, so the natural wood color isn't permanent there.

Push fiberglass insulation into any cracks around door- and window jambs with a putty knife or scraper to cut out drafts before you install trim. Start the trim about ¼ inch from the inside of the jamb and set galvanized finish nails with a nail set.

I install trim before I chink, so the mortar can fit tight to it. If you chink beforehand, you'll find there are spaces behind your trim you'll want to fill. Cut wire mesh and force it into these cracks, then push mortar mix in and smooth against the edge of the trim. You can use steel wool to take off any mortar stains later.

Doors and windows are a very important example of your craftsmanship. Very often the impression of the whole house is created in details such as window configuration and trim. Spend time on and give attention to these highly visible details.

One of my favorite front doors, of recycled heart white pine from about 1840. The hinge design repeats the arches on the five dormer windows on this large house we built in Charlottesville, Virginia.

CHAPTER ELEVEN

Chinking

IF THE LOG STRUCTURE fell out of favor for a single reason, it was the tedious chore of constantly maintaining and replacing the chinking that filled the spaces between the logs. Chinking filled in between log surfaces that weren't even or straight, and it supposedly kept out drafts, rain, and small creatures. The "chinks" themselves were thin splits of wood or thin stones wedged diagonally into the spaces to provide backing for mud, clay, or lime-mortar veneer.

A Little History

Traditional chinking never did work very well. Wood swells and shrinks with changes in humidity because of the cells' tendency to hold water. Lengthwise movement is negligible, but thickness and width change considerably, and this movement breaks up whatever chinking remains after winter freezes and summer rains.

Chinking is vital to the hewn-log construction that's traditional across most of America. This isn't as true of the round-log techniques that are more common in Canada and the western United States. In some kinds of round-log construction, which draw on Scandinavian heritage, each log is scribe-fitted to the one below, leaving a smaller or nonexistent chinking space. Still, drafts were an accepted fact of winter life.

But the American pioneers who settled in the Appalachian valleys 200 to 300 years ago didn't have the skill, the material, or, in many cases, the time for this painstaking fitting. Building shelter had to be fitted in between clearing land and getting in crops, and

This entire house had to be rechinked. The cracks between the logs had to be widened and the edges squared to leave a sharp corner on the bottom of each log. The new chinking sloped outward to shed rainwater. Chinking has a twofold purpose: to keep out the elements and critters and to help the logs shed water. With proper chinking, a log house will last many generations. Without it, the house is doomed.

that meant it had to go quickly. A gathering of pioneers could hew out a cabin and put it up in a day.

With the current revival of log building, filling the spaces between logs is once again a major problem. Log kit manufacturers have devised all manner of splines, locking grooves, and plastic gaskets to substitute for the old-style chinking. Some of these innovations may prove to work well over the years; others have already fallen by the wayside. Because most of these kit logs have not stopped "moving" when assembled, they have been known to twist out of the clever spline configurations.

Either way, I've got a strong prejudice against factory-made, fit-together log kits. They are expensive, and, to me, they aren't the real thing. Building or restoring your own traditional hewn-log house lets you participate in history, in a sense, and it can save you a lot of money.

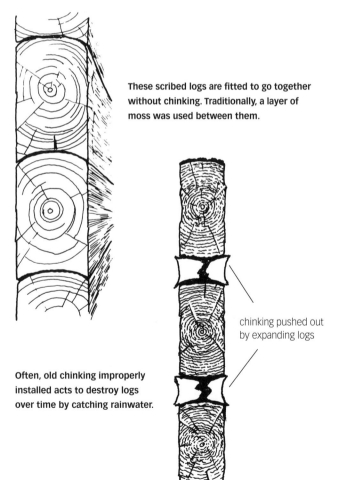

These scribed logs are fitted to go together without chinking. Traditionally, a layer of moss was used between them.

chinking pushed out by expanding logs

Often, old chinking improperly installed acts to destroy logs over time by catching rainwater.

Finding Out What Works

Modern materials and techniques that last but look like traditional chinking are the key here, because they solve the age-old problem of permanence.

I grew up around log cabins, and that meant trying different ways to make the chinking more effective and permanent. When I was 11, my father, older brother, and I tore down, moved, and reassembled a log house that had provided little real shelter because the chinking was constantly falling out. The logs had been notched so that they were close together, and the voids had been filled with solid concrete mortar. When the logs expanded during the wet season, the chinking broke up like ice on a pond.

We shortened and renotched the logs for a wider crack. Before we filled the space with a lime-based mortar, we drove rows of nails across the spaces and strung barbed wire in there. This gave the chinking some reinforcement. And it is still intact almost 60 years later.

But still those logs shrank with dry weather, opening cracks next to the chinking. Caulking seemed the ultimate answer, but it was expensive and required periodic maintenance. We also questioned the sense of using stone shards and wood splits behind the chinking mortar because they didn't seem to do much to keep out the wind or cold. But neither did filling the entire space with the masonry mortar.

As the years passed, I graduated from barbed wire to chicken wire, then to hardware cloth, and finally to metal lath, which is designed to hold plaster. I also tried various kinds of insulation, but it wasn't much good at keeping out drafts. Sprayed foam was better, but messy and hard to contain as it formed. It's expensive too. The best insulation that I've found is fiberglass batting. If it's packed more tightly than usual so it springs outward against the logs, it really keeps the wind out. It's not as efficient that way as insulation, but it does work. It takes about two times the quantity you'd use for normal insulating.

I ultimately settled on the following routine to chink my log structures. With a chain saw, we cut a one-inch-deep groove, a little over one inch in from the outside face of the log. This is only on the under-

Here are examples of why chinking in cabins in previous eras failed. Brick and stone "chinks" smeared with mud or concrete moved with the change in weather and allowed rain to damage the logs. The aim today is to re-engineer the chinking to make sure that rain doesn't get inside.

side of each log, and it's for the strip of reinforcing wire lath to go up into. When we tried shaping this top edge of the lath and nailing it in at the tucked-in angle to the irregular log underside, we lost a lot of time. Nobody wanted to do wire-lath duty. Next, I cut the strip of metal lath with manual or electric tin snips. It should be a little wider than the space and should be fitted flat up into the groove and slope out from top to bottom.

The top edge, being straight, goes into the groove no matter how irregular that log surface is, and the slack is taken up. At the bottom, we trim with tin snips to fit the lower log surface, and nail the lath with six-

penny nails about half an inch in from the face of the log beneath.

I usually parge the exterior first (after the doors, windows, and gable ends are in) to get the house weather-tight. Plumbing and electrical can be run in between the logs on the interior before doing that side. Fiberglass batting is stuffed into the crack to fill it to within about one inch of the inner wall surface. I buy the paper-backed rolls, which can be divided into strips of the right width with a handsaw by cutting through the roll. (This is like slicing an ice-cream roll.) Then I strip off the paper to get a tight fit. I favor six-inch-thick rolls.

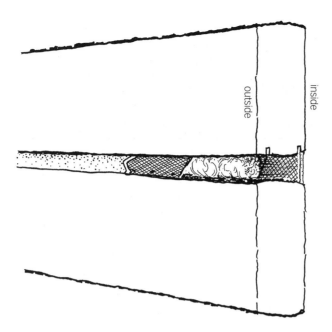

Pack fiberglass insulation against the exterior wire lath. The interior lath can be nailed in more vertically than the exterior, since the interior does not need to shed water. The insulation is sandwiched between the mortared laths. Both mesh and mortar should be sloped out at the bottom to drain water away from the log below.

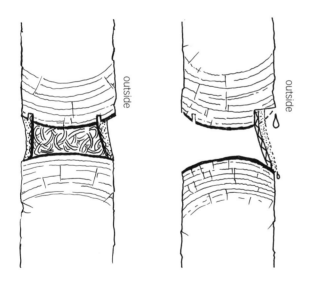

Chinking must be tucked under the lower corner of the logs so rain won't get behind it. If a log is rounded, cut a notch in it to leave a corner for rain to drip off. It is less critical for interior chinking to be sloped; in fact, some people prefer a more vertical, smooth look to an interior wall. It is the fiberglass packed between the two layers of chinking that provides the insulation, keeps out the wind, and "moves" with the log movement as the weather changes.

After the house is closed in and the electric wiring is done, a second strip of lath is cut to fit in the same manner and applied between logs from inside the house. The inside lathing need not be sloped, but it's easier to trowel the mortar onto it if it is.

The next step is parging cement mortar about half an inch thick onto the metal lath on each side. Then, when it's dry, we apply two coats of exterior-grade masonry sealer to the outside.

Preparation

About chinking green logs — don't. Although it might seem like the drying logs will settle and the chinking will stay tight, it won't. The log face is wider than the notch, so it shrinks more than the notch lets it settle, and the crack opens. The fiberglass inside expands so there's no space. And even without the caulk, there won't be a draft but a crack will open.

Any green logs, whether in new construction or as replacements, must season at least a year before chinking. You can fit the lath between the logs right away and then plaster later on if you like, but the lath is liable to rust unless you paid extra for the galvanized type. It's also important to make sure the logs aren't heavily rain soaked before chinking. Summer and fall are the best times to chink, when the weather is dry.

Allowing the logs to season is important. A 12-inch-wide log can shrink as much as half an inch. That's four inches for an eight-foot-high wall. So the whole house shrinks down as it seasons. If verticals like door and window facings, interior stud walls, and stair supports are installed before this seasoning takes place, then they either buckle or prevent the logs from settling in one area, which opens larger cracks below.

This brings up the question of how wide the chinking cracks should be. Early builders tried to leave as narrow a space as possible between logs to minimize chinking (unless the house was to be sided over immediately, as many were). As my first experience showed, this doesn't work very well. I like two inches of space as an absolute minimum, but four inches is better and looks better with wider logs. I restored a house in Kentucky with some whopping 10-inch cracks between logs and one in Virginia that used whole bricks up on

Cut the metal lath to fit the irregular chinking width of each space between the logs. Metal lath sheets permit you to fit spaces that are either very narrow or very wide. Install the metal at an angle, sloping outward. Nail into both the upper and lower log. Since this was an on-site restoration, there is no groove in the upper log.

end as nogging. But it does get more difficult to keep the wet plaster on lath wider than eight inches. And you don't want to see more chink than log. Nevertheless, you can alter the width between the logs on your cabin only so much. If your logs are small and narrow, you cannot change that. If your chink joint is wide, you can't change that either. You will need to play the hand you are dealt, so correct installation of the chinking wire will be your path to a workable chinking job.

If an old wall has narrow spaces, often there is deteriorating sapwood at the top and bottom of the logs where they were not hewn. This punky sapwood should be removed with a chain saw or hatchet to widen the spaces between logs. Another way to widen the gaps is to insert spacers into the notches.

Installing the Lath

Cutting metal lath is my least favorite job. It'll tear your hands apart if you don't wear gloves. Use sharp tin snips to slice up the two-by-eight-foot sheets and try to cut between the lengthwise diamond-shaped spaces, so you won't leave long, wicked spines. We use

CHINKING MORTAR-MIX FORMULA

1 part lime

2 parts Portland cement

9 parts sand (redbrick sand gives the chinking a soft gray/tan color)

Mix together all ingredients. Add water until the mix reaches a stiffconsistency. Trowel about ¾ inch thick over metal lath, installed at an outward-sloping angle.

Do not use premixed masonry mortar (intended for brick or stonework) for chinking, which is a thin, free-standing application. Do not use pigment to tint the mortar.

Mortar is troweled onto the tucked-under slope of the wire lath. The mortar must be pressed up under the top log and into the bottom log to bond with the wood of the logs. If the lath is installed properly, the outside mortar chinking will protect both the log and the chinking surface from rain and wind.

After the mortar has cured, it is waterproofed with two coats of masonry sealer. Sealing exterior chinking is crucial. Sealing interior chinking keeps down dust but is optional. After a season of heating the house, caulk any cracks with high-silicone caulk.

electric snips for this first cut. After you get a near fit, push the piece up into the groove you cut or start nailing it into the crack, then snip to an exact fit around knots and dips at the bottom. Don't force the stuff in. A common mistake is to let it curve inward, making it almost impossible to do a good plastering job. The lath should be rigid, flat, and sloped. Even inside you'll find it easier to plaster if you can tuck the mortar under at the top and slide the trowel over the bottom log corner as you do outside for weatherproofing. Nail the lath every six inches or so with sixpenny nails to hold it rigid.

Applying the Mortar

There are a lot of different mixes, which combine everything from clay to lime and horsehair to fiberglass. But unless I am doing an authentic restoration for a museum, or National Historic Register work, I use a mix that's a lot like masonry mortar. I use one part lime, two parts Portland cement, and nine parts sand, mixed to a stiff consistency. I *never* add pigment to chinking plaster. A friend of mine was given free dyed masonry mortar and ended up with bubble gum–pink chinking. At least he takes the ribbing we

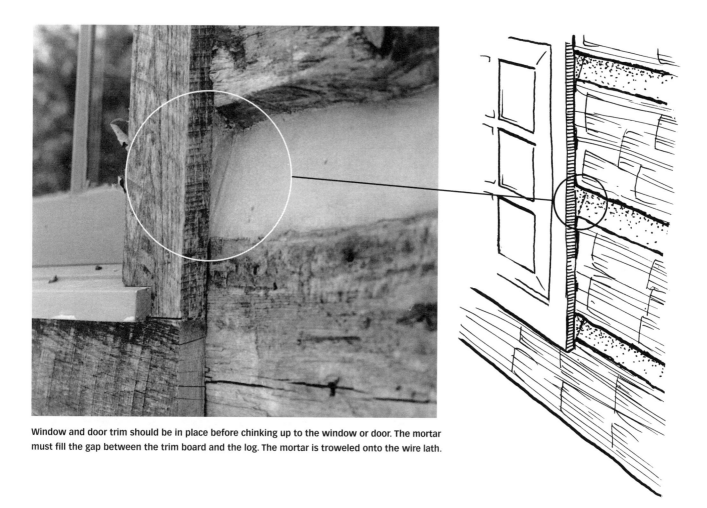

Window and door trim should be in place before chinking up to the window or door. The mortar must fill the gap between the trim board and the log. The mortar is troweled onto the wire lath.

all give him with good humor. I do use redbrick-colored sand to get away from the blinding white plaster color, though. This gives me a light gray-tan, and a more nearly authentic look.

I have not had good luck with premixed mortar, which you can find in any mass-market building-products store. This is intended to be used between bricks or concrete blocks. It is not intended to be applied in a thin, vertical layer that will be subjected to the harsh elements. I find that use of this mortar results in a great deal of cracking and flaking.

To apply the plaster, first dampen the log surface with a water-filled brush or spray, and then lay on the mix with a pointing trowel. If the lath is properly applied, no scratch coat is necessary. Start the trowel up under the corner left by the hewing of the top log and slope the mortar down so that it comes out flush at the bottom to shed water.

This tucked-in slope is such a simple matter it should be obvious, but apparently it isn't. I rechink

This photograph shows how water falls off the lower edge of each log and is shed by the chinking rather than working its way inside the log and the house. We use two coats of masonry sealer to waterproof the mortar.

We were called in on a consultation for a cabin that had been chinked with a plastic-based product seven years previously. These products are installed in a ¼-inch-thin application over a backing board or foam board stuffed into the chinking crack. In this case, the plastic mortar had peeled off like a Band-Aid in dozens of places on the cabin, both outside and in. We recommended that all the plastic mortar and backing boards be removed and that the whole house be rechinked with wire, mortar, and fiberglass.

several log houses a year that were done wrong — some very recently. You see, if the mortar is flush at the top, when the log shrinks it will open a crack that funnels rainwater right into the space. When the log expands, it pushes the chinking out farther. Some water runs inside, some outside, and a lot of it just sits there to rot the wood.

The plaster should be applied as dry as it can be easily worked, or it will shrink as it dries and cracks will appear. But after it's on, it should be kept moist to allow it to cure properly. In dry weather, I wait a couple of hours, then spray with a mist of water. In direct summer sunlight, I cover it with plastic sheeting for a couple of days, and lift the sheeting to spray it every two hours with water while the sun remains on it.

Special problems can arise. Once we restored a pine log house that got periodic southwest wind–driven rain. The old wood absorbed water every possible way — through checks and the tiny holes left by powderpost beetles, and even under the edges of our sloped chinking plaster. We finally had to caulk all the cracks outside with clear caulk (none of the available colors matched) and to paint the wood with a combination of creosote, oil, and bleach oil to retain the gray color. Since then I've always sloped the chinking in more at the top on the weather sides of the house, and I've always cut away any deteriorated, spongy wood that can absorb water, leaving a sharp corner at the log's bottom edge to tuck the chinking under.

At window and door facings and at corner notches, I stuff thin cracks with fiberglass and use a mason's pointing tool to push plaster in over it. Any application of cement plaster should have metal reinforcing in it, even if it's only a nail in a narrow crack. If the crack is a half inch or less, use caulking.

As waterproofing, we apply two coats of the clear exterior masonry sealer to the cement chinking as soon as it's dry — about a week later. This stuff doesn't shine, and soaks in quickly, so you can go right over the first coat with the second.

Now, after a couple of months of winter heat in the house, the moisture content will go down and the logs will shrink even more. Then you'll notice a thin crack opening at the top of the cement. Wait until this happens, then run a bead of clear caulk into it. Modern caulks last a long time, and if it can get into the crack, it'll flex with the wood and stay tight.

There are lots of synthetic chinking compounds on the market, but none of them has been around long enough for us to really know about its long-term effectiveness. Most of them are petroleum-based plastic, which breaks down in ultraviolet sunlight. They are also all very expensive. We use a high-silicone product that weathers well.

Scribing logs and shaping them further so as to allow chinkless fitting in the Scandinavian method does work. B. Allen Mackie, in British Columbia, is the acknowledged authority on this method. It's used with success by some of the current builders, but it requires more logs in a house. It is also quite time consuming (and therefore costs more money).

Using a plastic-based product means you'll have to rechink often. Maintenance is something you generally want to avoid in a log house, so don't build it in with the chinking.

CHINKING CHECKLIST

☐ Install window and door trim before chinking, if possible, for a tighter fit.

☐ Don't chink green logs.

☐ Allow a minimum two-inch chinking space between logs, but four inches is better. Trim back logs if necessary.

☐ Install metal lath at an outward-sloping angle, cutting with tin snips to fit the irregularities of the log. If there is a groove in the upper log, push the lath into it. If not, nail the lath an inch to 1½ inches under the upper log and slope the mesh toward the outer edge of the lower log, leaving half an inch for mortar.

☐ Dampen log surface with water.

☐ Lay on chinking mix with trowel to half an inch thick.

Log house chinking should be a natural complement to the logs and not call too much attention to itself. We use brick-red sand to take some of the brightness out of our chinking. This gives a soft tan — rather than cement gray — color to the chinking.

CHAPTER TWELVE
Floors and Stairs

POOR FOLKS LIVED ON DIRT. And so did lots of other, more solvent pioneers — at least temporarily. They had no choice during those first hurried seasons of clearing, getting seed into the earth, and moving out of the wagons that had been home for months.

A dirt floor meant the sill logs were also usually on the ground, so these cabins didn't last long. I have not seen a hewn-log house with a dirt floor, and only a few round-log huts, in the wearing-out days of the Depression, laid on dirt.

I remember temporary camps and communes built during the 1970s in which the buildings were set on dirt for a while, but "back-to-the-earth" movements were hardly authentically pioneer.

Watered and tamped periodically, dirt served well enough, apparently, because many of our ancestors seem to have survived on it. As soon as possible, however, something more permanent and satisfactory was added.

Puncheon Floors

The puncheon floor in its pure form was round logs split down the center, with each half laid in the ground, flat-side up. The splintery surfaces were worked with an adze to give a rough but very solid floor. Sometimes this was rubbed with stones and sand to smooth it further.

Again, the earth did its work on these timbers, and they did not last. So, often the puncheons were raised, becoming a sort of continuous expanse of floor joists, with ends wedged between the sills and next logs up. These raised puncheons were fitted more carefully at

Staircase at Page Meadows. Basement stairs are "stacked" beneath the main staircase. John Beard, of Free Union, Virginia, made the railings. The floors in the house were in mixed condition. Where possible, we kept the original, resanding and waxing. Where too deteriorated, we replaced it with remilled old heart pine.

their edges because drafts and small creatures could enter there. Typically, these raised puncheons were installed green, with a handsaw cut run between them for a close-fitting trim. As they shrank, these were moved together and another puncheon installed to take up the space.

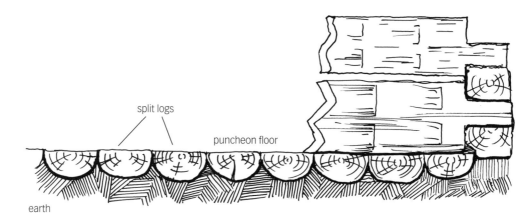

split logs

puncheon floor

earth

Puncheons and logs were laid on the ground in early cabins. They always rotted.

Board Floors

Pine boards and logs became the favorite flooring and building material among the settlers, where available. A wide pine floor is traditional in older houses, sometimes shiplap or t.i.g. to ensure a tight joint even with shrinkage. Oak was durable, but it was harder to cut and difficult to lay. Generally the downstairs floor was six-inch boards, narrow to use edge grain for toughness. The upstairs floors were often wider and made of softer wood because there was less wear.

Given many decades of scrubbing and the wear of countless feet, pine takes on a rounded mellowness all its own. Wide pine flooring is highly prized by restorers of old houses. I use remilled or recycled old heart pine t.i.g. in just about every house I build.

The flooring was traditionally laid directly onto the joists, with no subfloor. Here was an intriguing phenomenon that I encountered often in old houses, and in the early years could only conjecture on. A series of holes was often bored into the top surfaces of the joists, usually about ¾ inch in diameter. I'd found these both downstairs and up, and puzzled over them for years. When dismantling just about every old house, there they were.

My guess was that the holes were for a pry bar, to tighten cracks as the floor was laid. Certainly I had labored long to this same purpose on every floor I'd installed, and holes would have made it easier. Since, I've learned that this was indeed standard practice.

Today's Choices

Flooring today offers so many choices. Choices for the subfloor: plywood if nothing will show; one-inch t.i.g. if it is to be visible overhead; two-inch t.i.g., which can become the combination ceiling below and the flooring of the upper floor. Of course, carpeting and sheet or tile vinyl over a subfloor have their uses — even in an "authentic" log house that is to be lived in today.

The most common flooring in a log house is a wood floor. Again, the choices are myriad. New commercial flooring — oak, birch, maple, pine — can be purchased from any lumber company. However, it is almost impossible to get new flooring in the wide widths so strongly associated with early log houses.

Several companies are providing remilled flooring in many kinds of wood; oak, pine, and maple are popular. Remilled flooring comes from lumber, which has been retrieved, recycled, and resawn from other sources such as beamwork from torn-down factories or old logs. You can often get this flooring in wide widths — 6 inches, 8 inches, and 10 inches are common. This flooring is more expensive but can "make" a room.

Then there is recycling an old floor, down to the last wear pattern. Removal of the old floor is tricky. It can splinter at the tongues or grooves (you pull the tongue up first). The patina on old, worn flooring is wonderfully warm, so if you restore, try to keep it as it is. If

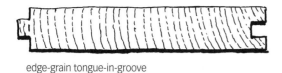

edge-grain tongue-in-groove

Tongue-in-groove flooring needs to be pried tight into place and nailed in the tongue.

flat-grain wide shiplap

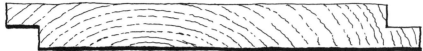

you reuse worn flooring, keep the pattern the same so thicker and thinner boards will blend.

Wood floor finishing can be as simple as scrubbing, or it can involve sanding, oiling, or waxing. Or, it can become as complicated as staining and varnishing with polyurethane or another polymer. Here, personal preference and the limits of your budget come into play.

The Joists

Let's begin flooring at the bottom. I space log or heavy beam floor joists 16 inches apart, often with any odd extra joist centered with the front and back doors to take the heavier traffic. Then I lay a subfloor of full one-inch rough-sawn oak or pine, usually insulating under it. Or we'll use plywood if there's to be no basement living space. Plywood doesn't look good overhead. Then a layer of tar paper goes on, and the floor is laid on that. Wide boards fastened with pegs are the most attractive, although square nails look good too. Tongue-in-groove should be nailed through the tongue with finish nails and a nail set. A nail gun can be rented from the lumber company or tool rental store to speed this operation. Wide boards will cup if nailed only in the tongue, so we face-nail there, setting the nails.

If you use conventional lumber for joists, use 2×10s or 2×12s, 16 inches on center, with blocking or criss-cross bracing at least every eight feet. The 2×10s will span up to 16 feet nicely, but I like 2×12s for longer spans. Smaller floor joists can be used with a girder or sleeper bracing them at midpoint, set on piers or a solid foundation.

You can use two-inch center-matched lumber for the ceiling/upstairs floor. If you use a subfloor instead, that subfloor will become the downstairs ceiling, so the underside needs to be attractive in its own right. Nail the flooring to the subfloor with nails short enough not to come through the underside between joists. Use longer nails at the joists.

I wanted to use some nice, 1840s, hand-planed t.i.g. pine paneling as the ceiling upstairs in our current house, but it was too thin for strength over our 42-inch-on-center beams. The loft space was to be used for children's bedrooms, with all those dormers for light, so we decided on a different approach. We used the paneling, unplaned-side up, over the beams, then

Log cabin class students lay a floor system at Bear Mountain Outdoor School near Hightown, Virginia. We routinely build a cabin, through the roof framing, in a weeklong course.

laid a plywood floor over that, fastening with short drywall screws to the paneling. Then we carpeted the strengthened ceiling/floor. (The subfloor was actually on top here.)

It's a bit difficult to finish off flooring at the log walls with molding, but if you smooth the worst humps out of the logs at floor level, you can get a strip of quarter-round to fit tight, with perhaps some caulking behind it. We scribe the last flooring board carefully and don't use any molding. We do use clear caulk.

An addition to our Missouri lean-to — containing a small office and the baby's room — was floored with recycled oak flooring, which we obtained for free. It was laid over a subfloor of rough-sawn oak on oak joists. Recycled flooring requires lots of sanding because the pieces aren't in their original wear patterns, and you get high and low spots against each other. But it has a glow unlike new wood flooring.

Our current house additions have beam joists with subfloor and remilled heart pine flooring. It's easier to concentrate plumbing in these more conventionally floored sections.

Slab Floors

In recent years, concrete slab floors have been used in log houses, and have been added to restored older cabins. Sometimes this is painted with masonry paint; sometimes it is left bare, to be covered here and there with rugs. Sometimes floor tile is used over it. A slab floor must always be vapor-sealed beneath to stop moisture from coming up. It also means a raised foundation, visible inside beneath the first logs or the wood is too close to the ground. Too often, the concrete is poured right against the logs with no vapor barrier, and that rots the logs.

In one restoration, the owner decided to use a low-ceilinged basement as an "English basement" (kitchen and eating area in basement). The floor needed to be lowered by at least eight inches. The only problem was that this 600-square-foot basement floor was a six-inch concrete slab. We had to jackhammer it out for days. Cleanup was almost as much work as the tear-out. Then, we dug down and graveled the full basement. A new slab was wired and poured, becoming the base for vinyl flooring.

This antique, recycled heart pine floor was laid and sanded in the usual manner. The owner engaged a friend to create a stenciled design to distinguish the dining area from the rest of the house and to match the shape of the custom-built table.

Dan McRaven lays insulation between floor joists. We often tack chicken wire or other mesh to keep the insulation from sagging. Some builders use wire "tiger teeth" bent up into the joist space. However, metal screening is more effective and more stable.

Flagstone

There are basically two ways to install a flagstone floor. The stones can be laid on a concrete slab or they can be bedded in sand over a filled floor space.

Both our Virginia house and our Missouri house had stone floors. In Missouri, we gathered it from the nearby creek bed, with some of the limestone as long as five feet and maybe two inches thick. It's a very natural and durable material, though it was not widely used by the early builders. A durable mortar is necessary between the stones; the scarcity of mortar early on may have been why few pioneers used stone.

I lay a vapor barrier of heavy plastic sheeting over four inches of gravel, which is over the raked and tamped earth, then cover this with sand. Each stone is then bedded flat into the sand and mortared between the edges. As in all stonework, it's a jigsaw puzzle. There's really no point to a concrete slab beneath a flagstone floor if moisture is sealed away from the base.

Flagstone can be sealed with masonry sealer, then scrubbed and waxed as often as desired. I will warn you that those beautifully flat stones you selected never seem as smooth when laid, but you get used to it. We did find that rigid furniture with more than three legs had to be relegated to other, non-flagstone floors in the house or wedged level in a permanent place. We used those flexible wood-and-canvas director's chairs, from necessity.

The Virginia log house section has heat pipe, a 400-foot coil under cut soapstone, laid in a similar fashion. Fill dirt, then gravel, went inside the raised foundation, then was covered with six-mil plastic sheeting. The pipe was laid on this and covered with the bedding sand, which held the polybutylene coil in place. We grouted the soapstone with masonry mortar and sealed it with two coats of masonry sealer. We wax it once a year or so.

Until recently, soapstone was available from reactivated quarries in central Virginia. As of this writing, however, it has become an expensive option, sold through stone yards by the square foot. When sealed and waxed, the black stone shows veins not unlike marble.

Today we build and restore log houses using a variety of flooring materials. The favorite in Virginia is recycled heart pine, left with the patina of age on it.

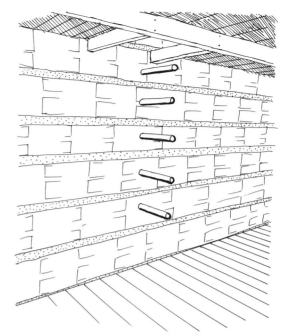

If there was no room for a stairwell, or if the loft was used just for storage or a place for the children to sleep, then early builders often installed a peg ladder into the log wall.

Often, we buy the remilled heart pine from old beams, which is kiln-dried and tongue-in-groove. Other choices are oak, maple, and even cherry flooring, all over a subfloor. Upstairs we often space the joists wider, using two-inch t.i.g. Sometimes we cover this with sound-deadening material, then carpet. Often it's the finished floor. Our non-soapstone floors are three-inch t.i.g. heart pine, recycled from an old cotton mill.

Whatever you do, build your floor strong. Too long a span between joists is no savings — nor is thin flooring or subflooring. It's embarrassing for your guests to fall through.

Stairs

You have a variety of stair choices to reach the upstairs or loft. Most are used traditionally. The classic one is a row of stout pegs out from the wall leading up to a hole in the ceiling. You can nail a ladder to the wall, but this won't meet building code either. Another option is one of those disappearing or folding ladders. Or you can build a steep staircase, at the end of the house always, in a one-and-a-half story cabin. If it's at the front or back of the house, you'll bump your head on the roof before you reach the top. Again, check

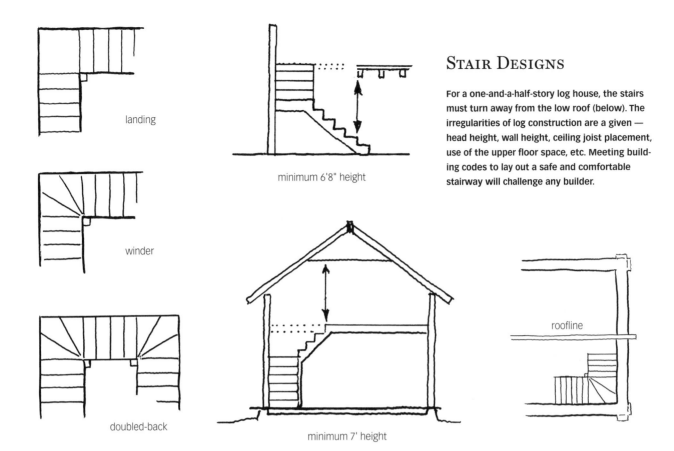

landing

winder

doubled-back

minimum 6'8" height

minimum 7' height

roofline

For a one-and-a-half-story log house, the stairs must turn away from the low roof (below). The irregularities of log construction are a given — head height, wall height, ceiling joist placement, use of the upper floor space, etc. Meeting building codes to lay out a safe and comfortable stairway will challenge any builder.

building code requirements for stairs in your area. Codes are requiring reduced riser height, so more steps may be needed for the same floor-to-floor height. In our opinion, however, shorter steps are safer, but narrower treads are not.

The problem here is the steep angle, because you must have room to get around and onto the steps at the bottom, off at the top, and up 11 or more in between. If you plan to move furniture up or down, prepare for a hernia. Codes require a minimum 9-inch tread and maximum 8-inch riser, 36 inches wide. You also need a 36-inch floor space (landing) at top and bottom before you start the stairs.

I build stairs with a sane pitch, made possible by starting up the front or back wall to a landing or winder well below head-bumping level, then up the end wall. It uses up a lot of living space, true, but leaves a logical place for a small coat closet downstairs. We have a half-bath under ours. Upstairs it just takes a lot of space. Try carrying a bathtub, wardrobe, chest of drawers, or other bulky mass up or down and you'll gladly sacrifice the footage. You'll do it more gladly as the years advance.

When laying out stairs, remember that you'll have one more riser than tread, because the upper floor is the final tread. Divide the total inches from finished floor to floor by the maximum 8-inch riser to start with, then adjust. You'll usually have 11 or 12 steps in the average cabin, maybe a little under 8 inches each. Then it's a good idea to draw out the treads and risers on the wall itself, including winding steps, if any. Don't forget to allow for tread thickness in this layout.

Choose your own stair construction. A favorite has always been simply treads nailed into angled cutouts in the side framing stringers. You may prefer treads laid onto nailer strips set inside the stringers with glue and screws. Another variation is the treads end-nailed through the framing, with risers for support.

Whenever I attach any horizontal member to a vertical — as with a stair tread, joist end, or porch rail — I like to mortise it. This was common before nails were plentiful, and though time consuming, it is stronger. So consider mortising each stair tread into the stringer, with perhaps glue and pegs to help. Stairs should not be fancy and out of place, but are a good way to display careful craftsmanship.

In one-and-a-half-story cabins, stairs must be placed so the top step ends up where there's headroom. You must have six feet eight inches of vertical clearance from the front of any tread to anything overhead (beam, ceiling, roof). This means having the stairs either end in the loft out in the center of the floor or at least head in the direction of the center, where the roof peak gives the headroom. We've also had to turn stairs back on themselves to fit them in, because of window placement or just not enough room. That often lets them come out where there's vertical room.

If the loft is for human occupancy, building codes have tight restrictions. If it's for storage, no problem. Also, if there is one staircase that meets code, a second doesn't have to. Where two log pens or an addition make a second staircase logical, this applies.

Railings are a must, and building codes require that they be 32 inches high with openings that are no wider than four inches. The only way I've avoided these code requirements is to be granted a variance in the case of authentic restoration of an original.

Window placement is a mess if you have a languid staircase taking up most of two walls. An 8-inch rise and 11-inch tread are all right, but anything more gradual and you have large dark areas where there should be windows, along with more wasted floor space.

Many houses we've dismantled had the entire staircase wedged between corners and the fireplaces,

Our back stairs are turned around a hewn post, which supports the headed-off beams above. This is how we get extra steps in along an outside wall to meet the building codes.

STAIR TYPES

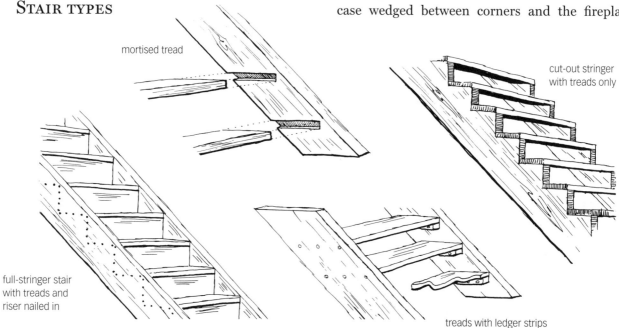

mortised tread

cut-out stringer with treads only

full-stringer stair with treads and riser nailed in

treads with ledger strips

which were close, sometimes five feet away. Of course there was no window in that half wall, and the treads were maybe only five inches deep. Like it or not, you went upstairs on tiptoes.

Circular stairs, around a central pole, are popular just now because they take up little room and meet most building codes. They can be bought for as little as $400 on up. Often you must cut and head off ceiling joists to accommodate these (and any other) stairs, so you'll want to plan ahead. Also keep in mind that it's very difficult to move large objects up or down tight spiral stairs.

Here I'll note that code requirements apply to living spaces and not to storage. You can use a ladder to get up to a storage loft and be legal, as long as you're not storing warm bodies up there. And, of course, after final inspection and approval, and issuance of an occupancy permit, people will do more to their houses.

Sometimes we can reuse a complete staircase, or parts of one. It is essential that each riser be the same height and each tread the same depth for safety. So if you cut down a staircase for reuse, you may have to modify the whole thing.

Floors and stairs are where we use really good-looking wood. The Virginia favorite is remilled, oiled heart pine. Stairs, along with their railings, and interior trim and cabinetry are showcases for fine workmanship. Take a lot of time, and a lot of pride in them.

FLOORS AND STAIRS CHECKLIST

☐ Choose plywood or tongue-in-groove for the subfloor.

☐ Look for remilled flooring from specialty suppliers.

☐ Finish wood floors according to your preference.

☐ Consider whether the floor for the upstairs will be the ceiling downstairs.

☐ Seal a concrete slab floor to prevent moisture intrusion.

☐ Lay flagstone on a concrete slab or embed stones in sand.

☐ Check building codes before planning stairs.

☐ Choose a form of stair construction.

☐ Consider mortising each stair tread into the stringer.

☐ Remember that railing height is 32 inches and spindles are maximum four inches apart.

LEFT **Stringer mortised to vertical post. Treads were also mortised into stringers for this oak stairway. My brother Dan was the carpenter.**

RIGHT **Rustic stair treads of half-logs in Gordon and Jane Dolton's Michigan cabin, built by her parents in the 1930s.**

CHAPTER THIRTEEN
Stone Fireplaces

I REFUSE TO BUILD A LOG HOUSE without a stone fireplace.

At best they are the poorest of heat sources, but that's hardly the point. A log house looks somehow naked without a chimney, whether of mud and sticks, brick, or stone, and the room inside the basic cabin just sort of radiates from the hearth.

Early on, of course, the women cooked at the fireplace, hanging pots from a crane or iron bar across it. As the wood cookstove appeared in the hills and was (usually) installed in the lean-to, the fireplace didn't have to go all year long. And when the cast-iron heater came, there were log houses built with no fireplaces.

The Fireplace Mystique

I suppose the mystique goes back to our very early use of fire as protection at the cave mouth. Certainly the open fire has historically been a source of cheer as well as warmth. And, of course, the flames were often the principal light source for the settlers, along with their bear fat and bayberry candles.

Until recently, not much had been done to make a fireplace heat well, and little is done even now. Basically you have a fire in an opening in the wall with a sloped back to reflect some heat, and a narrow chimney throat opening to a wider bell over a smoke shelf to foil downdrafts. Maybe there's some firebrick as lining, or even a metal box inside, and a damper.

The dual-wall heat box in wide use now simply draws cold air from below, heats it, and allows the smokeless air to circulate behind the firebox and

The chimney on the Beaver Jim Villines cabin at Ponca, Arkansas, before restoration. This is drystone work, using the excellent local sandstone.

either convect naturally or be blown out into the room. That helps a lot, but it tends to rust out in a few years.

But still, in order for the fire to burn, large quantities of air must come from somewhere to feed the flames and then go on up the chimney to be wasted. This air has always come through cracks around windows, doors, chinking, between shingles, and up through open places in the floor. Eliminating these air sources would make it just about impossible to get a fire going.

One of the brightest ideas extant is a duct for outside air to feed the flames, to be closed when there's no fire. The air still goes up the chimney, but warm air isn't drawn out of the room and drafts are eliminated. I have used either a direct or a "Y" passage in concrete beneath the firebrick in some fireplaces, with screening over the outside openings to discourage varmints. The air comes in right in front of the fire and does its thing nicely. A device to keep ashes out of it is a simple necessity.

Another improvement using the dual-wall box, puts the cold-air inlets ducted under the floor to far corners of the house. This draws cold in, leaving low-pressure areas, which in turn are filled with the warm airflow from the heat box outlets. If no blowers are used, these ducts should be the full size of the inlets themselves. With blowers, you can use smaller ducts because the air travels faster. Without these ducts, the circulating heat box pulls air from the room very near the fireplace, and doesn't do much for the distant parts of the house.

A fireplace may be flush with the inside wall, with the structure completely outside. Or it may break into the room and be almost flush against the outside wall. Traditionally, it was largely outside, taking less room space and being simpler to construct. But that meant much of the heat was convected outside. I have built raised-hearth, stone-to-ceiling, floor-level hearth, heat box, and just plain fireplaces. I make some compromise with history here, because the old fireplaces did such a wretched job of heating. If you plan to construct a fireplace, get exact specifications; it is so easy to foul up the joy.

Catted Chimneys

I have seen, in my youth, mud-and-stick chimneys, but haven't built an operable one. Nancy McDonough, in her delightful book *Garden Sass*, notes the use of mud chimneys largely in the Ouachita Mountains of south Arkansas, and stone was almost exclusive in the Ozarks, the Alleghenys, and the Smokies. Henry Glassie also notes the heavy concentration of "catted" chimneys in the Ouachitas, which, curiously, is the home of some of the best building stone in the country. This may be simply a result of tradition — the

The Wintergreen fireplace is built entirely inside the house walls, which keeps heat, convected through the masonry, inside the house. Building codes require that chimneys be two inches from structural framing (in this case, logs), a dilemma for all builders.

building customs of a group of settlers of similar backgrounds became the custom of their children, and therefore of later settlers.

The catted chimneys with their mud "cats" and neat mini-log pens are admittedly works of art, but I will stay with stone.

Masonry Chimneys

In the East and red-clay South, bricks were used for chimneys. In the mid-Atlantic and into Tennessee and Alabama, a combination of stone and brick was common. Stone was used as a base, because the low-fire brick often softened in the ground. Also, bricks were easier to carry up high on the chimney.

Today, to save money, builders are using concrete blocks and parging over them with a kind of stucco. Stone or brick veneer over block is also common, but in many cases is really more expensive because two chimneys are actually being built.

This drystone chimney shows how a wider stone base was laid in the ground to support the weight. It is near the Mulberry River in Arkansas.

Traditionally, a fireplace opening was sawed in the end of the log house. The logs ends were faced with boards, as with the windows and doors, to hold the ends in line.

Chimney Foundations

After an opening was sawn in the wall of the house, large slabs of stone were laid on or in the ground, about twice the size of the hearth/chimney area. These spread the tons of weight over a wider area and minimized settling, which always happened anyway. The chimney settled at a different rate from the house, so chimneys weren't attached to walls. Over the years a crack often opened between the lintel stone inside and the first spanner log above it, and this was stuffed with some sort of chinking from time to time.

The ground was wetter and softer away from the shelter of the house, so the outer side of the chimney base often sank lower than the side near the foundation, and the chimney often tilted away from the house at the top. Most unrestored older houses show this, no matter how good the stonework. Sometimes a metal band made of old wagon tire was bent around the chimney and bolted to the house wall to keep the chimney straight.

Today we dig below frost to firm subsoil, then lay a reinforced concrete slab at least one foot thick and, at the very least, twice the area of the chimney base to start things on. For a 3-by-5-foot chimney base I dig a hole about 5 by 7 feet and at least 18 inches deep. Then I pour 4 inches of concrete: cement, sand, and gravel, mixed 1:2:3. Over this goes a grid of reinforcing rod, half an inch or thicker, about every 6 inches each way, almost to the edges. Then another 4 inches of concrete, another grid of rod, and a final 4 inches of concrete.

That's 35 cubic feet of mix, well over a cubic yard, and it means about a pickup truck full of sand and gravel and 7 sacks of cement. It also means you'll need well over 200 feet of rod. (The ends don't reach all the way to the edges.)

That's the pad or footing. You see, you'll have truckloads of stone, weighting on a C-shaped area about nine square feet total. The idea of the reinforced slab is to spread the weight, and also to keep the C-shape from breaking its way down through the slab. Concrete without reinforcing isn't very strong; it's not even as strong as dense stone. With steel inside it, it will do amazing things, like span a ceiling as beams or become the hull of a boat.

Lay stone in mortar on the slab, just as you did on your foundation footings. Build up solidly to floor height or leave ducts for fresh air or an ash dump.

You may have built a pier to support the floor joist sleeper end already (if any), or you may want to build the whole fireplace first to floor level. I lay the slab, then go up almost to floor level first, incorporating the sleeper pier if any, with flashing separating it from the masonry. Then I finish the fireplace after the house is up and pretty well settled. Or I do the whole thing at a later date.

Extend the slab well into the room for hearth footing. It can be less massive here, because it doesn't support the chimney. The hearth must extend at least 20 inches into the room, beyond the stone face of the fireplace. I bring up stone, through a hole left in the floor, to hearth level, then lay flagstones here for the hearth into the fireplace, at which point I switch to firebrick.

The Firebox

If you don't plan to squeeze extra heat from a metal heat box, build entirely of masonry. Bring the back of the fireplace cavity up and forward with firebrick in a slope or a curve to reflect heat outward. This back should stop only about 10 inches from the front stone wall, which is the lintel stone, and several inches (9 or 10 inches) above its lower edge. Here is where a

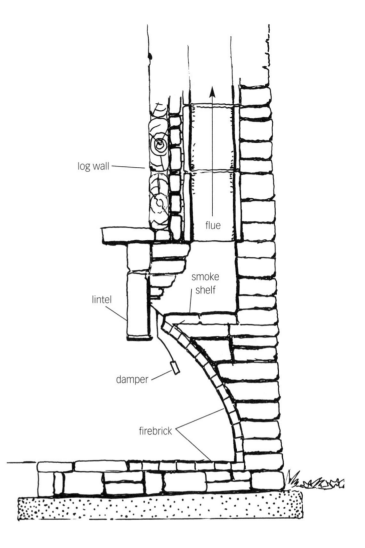

Cutaway side view of the fireplace. The smoke shelf is extremely important to eliminate smoking. Flue tile is a building code requirement. The log wall above the mantel must be separated from the flue opening with eight inches of masonry. The bottom edge of the lintel must be at least nine inches below the damper.

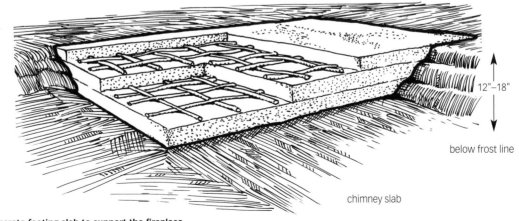

This is the reinforced concrete footing slab to support the fireplace and chimney. It should be at least one inch thick and twice the area of the chimney it is to support.

Logs and stones are irregular building materials, and joining them requires ingenuity. Trim boards and drywall need to be scribed to the logs or stones. Here mason Eric Bolton did an excellent job scribing a trim board to his fireplace.

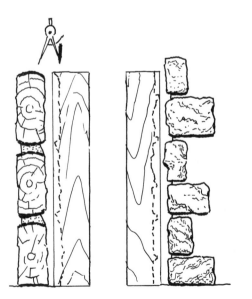

A good way to make scribe marks is to use a common compass. Run the tool along the rough surface and mark the board. You can then cut and trim and rasp the edge to fit the irregular surface.

damper, if any, will go. Dampers are useful to shut out the cold when there's no fire. If you buy a premade damper, shape this firebrick box to it. Remember to leave a ¼-inch space at the edges of the metal for heat expansion. Lay the firebrick with mortar that has fireclay in it (about half the lime content).

Above the damper, drop back clear to the back, leaving a smoke shelf to prevent downward gusts. Above this, narrow the chimney front to back and side to side to about 12 inches square for a 36-inch fireplace and go for the sky, lining with clay flue tile. Set the tile on this narrowed pyramid. The smokeless-room theory works a lot like the venturi in a carburetor, constricting the smoke flow at the damper, front passage, then opening out. Downdrafts reach the belled-out space above the smoke shelf and lose their sense of direction. Heat rising overcomes the reduced downflow here.

Line the inside of your fireplace cavity with firebrick, an expensive but durable substance that keeps annoying bits of hot stone from popping off and ricocheting around the room. Bare stone will also crack completely apart in a raging fire, so don't take the chance. There's a special cement, called refractory cement, for bonding this brick. (Or you can mix fireclay into the mortar.) I build the firebrick shape either first or as I lay the stone, using corrugated masonry ties to hold brick to stone.

The smoke shelf chamber should rise from the firebrick box and a steel brace across the front (leave expansion room) to step in and accommodate the flue tile. You can use brick here, parged inside with mortar with fireclay mixed in. Concrete bricks also work well here. Either way, be sure to leave room for the damper to function properly.

The lintel stone should, of course, come down below the top of the sloped or curved back so smoke will be more inclined to lift off. Otherwise, it will roll out into the room. This stone should be set on piers built up even with or inside the walls. I break the inside stonework out into the room from six inches to a foot, enclosing the log ends. Building codes sometimes require you to keep two inches away from the log ends here, so you have to do a vertical chinking joint over wire lath to fill the space or scribe a trim board to the stone and logs.

Lintel Stone

Look long and hard for the right lintel stone. A single, massive rock spanning the fireplace opening is a thing of joy and awe. It may be arched or flat; it may or may not be supported with steel. I use no support for an arch, and a heavy piece of angle iron for the straight span. Neither is really necessary to support a dense stone with some height to it. The settlers often used old iron wagon tires for support, because the big fires needed to heat their drafty cabins sometimes cracked the lintel stone. Settling, too, took its toll on the lintel. If you use steel bracing, allow ¼ inch or so of space at each end for heat expansion.

Remembering that this stone may be four feet or so long and weigh 400 pounds easily, get some friends to help lay it. We slide the really big ones up a heavy plank that is supported at the fireplace by blocks and a hydraulic jack. When it's in place, we let the jack down slowly, bedding the stone in its mortar.

For the Morrises' Possum Creek cabin near Charlottesville, Virginia, seven of us carried the lintel into the house and set it.

If your front stonework is one foot deep and you can locate only a thin, four-inch stone, lay it as a facing with other stone and concrete behind and supported by the steel. Try to use a stone at least one foot high, if you can find it. Height means strength in a span, much more than thickness.

Lacking a single stone, you may elect to keystone the span, either flat or arched. All this requires is that you build out from each side, on either a temporary or a permanent support, sloping each stone back. Fit the keystone in the middle. It will have to compress for the span to fall. Stone doesn't compress easily, and that's what keeps the span up.

The Flue and Chimney

Above the smoke chamber, use tile (codes require this) of a minimum 12 by 12 inches for a 36-inch-wide fireplace. This makes the chimney easier to keep clean and supposedly lets the smoke rise faster. It can be set on the smoke chamber you built up or on angle iron mortared into the chimney. Some masons leave space between the tile and the stone, filling it with dry gravel

The keystone in our arched fireplace was recycled from a West Virginia log cabin. The middle arch stones on each side are Dublin, Ireland, cobblestones that were used as ship ballast, and a gift from our friend Bill Cameron. The fresh-air-duct opening in front keeps the fireplace from drawing heated air from the room.

These drawings show both the stepped and sloped chimney shoulders, which are traditional above the fireplace. East Coast chimneys normally incorporated an upstairs fireplace, so this shoulder was at gable height.

Using the come-along to lift a large stone. The roof cutout will be flashed where the chimney goes through.

to absorb expansion. Others leave air between. Still others lay the stone right against it, ignoring expansion. I use tile, bringing the stone close, filling the space here and there with gravel. Seal around the top with masonry mortar, so rain won't sog everything.

Step the chimney in a foot or so on each side above the large chamber. I usually do it in two or three steps, for 45-degree shoulders here, or slope the shoulders with large slabs of stone.

At the roofline, whether inside or outside the house, you'll need to flash to keep out leaks and prevent wood decay. Set aluminum, galvanized or copper flashing into the horizontal joints in the stonework, and bend it down over counter-flashing up from the roofing.

At the ridge, a piece will have to be fabricated, so it can be set in stone and then cut, bent more than one way, and soldered.

Near the top, a bead was traditionally stepped out. Sometimes this was only a thin stone shelf, which was occasionally several graduated steps out, then back in. Earlier, it was believed that this helped deflect wind upward, but it was mostly for decoration. A bead is sometimes an indication of the age of a chimney, as is the mortar used. Wide, heavy beads were common in the 1600s, then reappeared in Victorian houses.

Of course mud or clay, or even drystone, was traditional in the backcountry, but lime mortar was used where available, and became more widely used after the Civil War. Modern cement mortar wasn't evident till around 1900 in the hills, about when the square nails disappeared.

I usually finish the outside work on a fireplace, then work the inside stone, mantel, heat-box ducts (if any), and hearth. Again, I like for the logs to be pretty well settled before I lay stone against them, inside or out.

Fireplaces take lots of time. You may contract yours out or even build your house without one and add it on later. Allow most of a summer if you do the job yourself, unless you know your masonry. I know a number of masons who regularly build a fireplace and chimney in a week. I don't. I like for each day's stonework to be thoroughly set up and cured before I pile the next thousand pounds or so of stone on. I aim for two or three vertical feet a day, with several days between, but you'll have lots of other things to keep yourself busy while this job is going on.

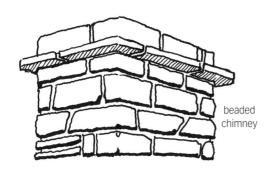

beaded chimney

The chimney bead is a stepped-out layer of stone that is purely ornamental. It is a nice touch for a log cabin.

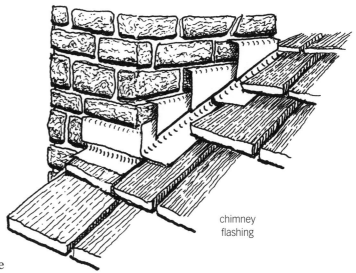

chimney flashing

Flashing is set into the stone of the chimney as it is built and bent down over counter-flashing up from the roofing. We use copper.

You'll need scaffolding when you get about to eye level. It can be rented or borrowed, or built of 2×4s as you go up. Anything cross-braced and strong will do, with planks laid across. A front loader is a delight to work with; you can stack stone, mortar, tools (and your feet) into its bucket as you work. They don't reach very high, though, so you'll eventually need something solid up in the sky. I usually rig a pulley and rope to pull up buckets full of materials, off the secured scaffold or from the ridge of the house.

You may, as I've said earlier, build the chimney and fireplace before you build the cabin. This makes it easy to get at, I'm sure, but I've never done it this way, and I don't know that any of the early builders did it. Again, I would be concerned about the logs settling as they shrink.

Woodstove

Wood heating stoves are a great deal easier to install. Even the modern Franklin variety requires only a hole in the roof, a multi-wall thimble to insulate the pipe where it goes through, a damper in the pipe, and some flashing on top. Double- or triple-wall pipe is required today in most locations to meet fire codes, at least where it goes through ceilings and roof. Put a cap on to keep rain out, and you're set. It is a good idea to get one of those metal sheets or a piece of slate to put under, to catch coals and sparks. Heating stoves are nice to cook stew on, too, and to heat water.

You can even heat a small cabin with a wood cookstove, which is quite versatile. But its firebox won't hold enough wood to burn through the night, and your indoor plants will likely freeze before dawn.

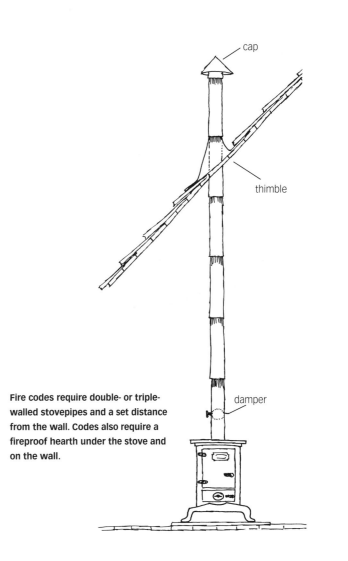

cap

thimble

damper

Fire codes require double- or triple-walled stovepipes and a set distance from the wall. Codes also require a fireproof hearth under the stove and on the wall.

The warming oven was a common feature in early fireplaces. This metal door I forged is for a kitchen fireplace at Page Meadows.

I think one of the neatest ways to heat a house is with a circulating woodstove that has a thermostat. A load of wood lasts all night, and the thermostat keeps the heat constant, so you don't have to adjust the damper all the time.

We have an 1892 cast-iron heater, which in past winters has gone up in several rooms where there was no fireplace. As a heat source, this stove put our fireplace, with all its improvements, to shame. But you can't see the embers and feel the cheer as well as you can at the open fire. I guess it's just the caveman in me.

STONE FIREPLACE CHECKLIST

☐ Lay a reinforced concrete slab as chimney foundation.

☐ Lay stone in mortar on the slab up to floor height.

☐ Extend the slab well into the interior for the hearth footing.

☐ Line the fireplace cavity with firebrick.

☐ Look long and hard for the right lintel stone.

☐ Use tile above the smoke chamber.

☐ Flash at the roofline to prevent leaks and decay.

☐ Keep structural masonry two inches away from logs or framing.

CHAPTER FOURTEEN

Porches and Additions

UNLESS YOU'RE A HERMIT, or plan to become one, scrap the plan for just the basic settler's one-room cabin. I don't care if you think you travel light, or if you want the house only for vacations or weekend stays, or for extremely intimate gatherings. One room just isn't enough, especially given the limits that log length and weight impose on the structure. Even a theoretical 40-footer, although possible, would shake every time a door slammed. It's just too far from bracing corner to corner.

So let's stick to the old basic 16- or 20-foot log pen, whether as a single or half a dogtrot or saddlebag, and add onto it. First off, I rarely build a log house without a porch — although the really early cabins didn't have them — and I usually build a lean-to. I will admit to a certain laxity in insisting on a loft, always to my regret. Everyone gets cramped for space eventually, and these additions provide the simplest, cheapest way to expand, just as they did historically.

But these days a lot of people want to leave the ceiling out so they can have a high, open space to the roof. Half the folks we build or restore for have this idea. Most of them change their minds after all the heat goes up there in winter, while their feet freeze down below.

We compromised in our house, leaving eight feet open to the roof over our 16-by-20, one-and-a-half-story kitchen addition, which is stone. My wife, Linda, has her office up there, with a railing at the edge. A big window washes light down from the gable into the kitchen. Dormers let light into the office, keeping it cheerful. And she can keep an eye on whatever is being plotted by the kids in the kitchen. A ceiling fan helps keep the air and heat where we want them.

Bill Cameron in a rare relaxed moment on the porch of our Missouri cabin. Bill helped with many of our building projects in the 1970s.

Additions in general are historically appropriate and logical. My favorite term for added-onto houses is the eastern Maryland and Virginia "little-house-big-house colonnade-and-kitchen."

The porch on two sides of the cabin for the McGees, which was built as a woodworking shop. It was later converted to a guesthouse — as have all our workshop cabins.

Porches

The settlers used the front porch for everything from drying herbs and produce to talking politics and courting. Assorted hounds basked here, ready in a flash to scramble full-throated after raccoons, peddlers, or chicken thieves. Harnesses and saddles always seemed to find their way here, even at well-to-do layouts with spacious barns. And the women would hang their shiny washtubs proudly on the porch wall, exactly as they later were to display their new washing machines.

And rocking chairs. Where else, in decent weather, would you have put Granny and her patchwork, your maiden aunt and her store of gossip, or your old uncle mumbling over his jug? Of course this furniture moved inside in nippy weather, but in the backcountry, rocking chairs were out more than they were in.

So you want a porch. It makes the house balance better; it keeps the rain from dumping on you while you fumble open the door, arms full of firewood, groceries, or somebody else's chickens. And it'll get just as hot inside as it did for the pioneers (every bit as hot if you cook on wood), and you'll need the air.

A porch is simply a heavy floor, usually raised, with a roof extending out from the main house, supported by two or more posts. Visually, the roofline of the porch is the part that's most important, so let's talk about that first.

Roofing the Porch

There were three basic ways the early builders went about roofing the porch. One was simply extending the main house rafters to seven feet or so from the ground, propping them with uprights, and shingling the whole roof together. This "cat-slide" is neat, with no break in the roof, making it easier to frame up and to shingle. There is a disadvantage, however, because with a steeply pitched roof on the main house, it means a shallow porch.

In another roof design, the porch rafters were attached to the house below the main eaves, which extended the porch roof at a flatter slope than that of the house roof. This allowed a deeper porch than the straight-line roof. It also left space for small windows to be installed in the upstairs wall, above the porch roof and below the main eaves. Dropping the porch roof below the main eaves gives less height to work with, however, and a deep porch still must have a sufficient roof pitch to shed water — say a 1/2½ or 1/3 pitch ratio for shakes.

I prefer still another arrangement — that of the porch roof joined to the main house roof, but at a flatter slope. This gives the maximum height at the house wall, allowing a deeper porch, which I like, still at a sufficient slope. Shingling the joint where the porch roof joins the main roof takes a bit of skill, but I underlay this with a strip of flashing. If you install a standing-seam porch roof, you'll crimp the vertical bends of the metal, then hammer them flat before seaming. We used to cut and solder here, but it's hard to avoid leaks when you do it that way.

The porch roof slope that looks best with your house is strictly a matter for your own eye, and as long as the shakes shed water, you have quite a bit of leeway. I attach the house end of a porch rafter to the end of a main rafter with one spike, letting the other end pivot and hang down temporarily. Next I get help, if necessary, to raise the porch rafter, swinging it up until I get the slope I like.

I usually frame up the porch roof at the same time as the main roof, so that I can go ahead with roofing. I brace it temporarily, and sometimes frame and floor the porch later. In pinning the rafter ends to the main house, I use wooden pegs or long spikes, such as gutter spikes. The same fastenings are used in pegging the other ends of the rafters to the long stringer. This stringer itself may be doubled and spliced 2×8s, because it should be the entire length of the porch, and you may have trouble finding a single timber or pole long enough. Usually we hew out a straight poplar pole for this.

Three porch roof styles. At left is the "cat-slide"; center is the broken roofline; and right is the separate porch roof, which is attached to the house wall using flashing.

Any time the edge of a roof is up against another surface (such as a chimney or a wall), you must install metal flashing along the incline. Copper is best because the color tones down and disappears. However, use the same metal throughout, since no unlike metals should touch or they will corrode. It is especially important to flash a porch roof where it attaches to the main house. Here, the roof was not flashed at all. Rain had been allowed to splash against the log and funnel down the log wall. This caused the logs to stay wet and to rot. Repairing this damage will be a major effort and expense.

Porches are a traditional way to add living space to a house. Open or screened porches on new log houses or restorations can be attractively designed.

If you've been efficient and already have the porch floor framed, you can spike or mortise the permanent uprights for the roof in as soon as the rafters are up. Or prop and brace the uprights in place, climb up, nail on the stringer, and lay the rafters that way. That's called planning ahead.

Building the Porch

Porches were often afterthoughts, so build yours any way you like. I like to slat the roof instead of decking it, and lay the shakes or metal with no tar paper under. The undersides of the shakes won't weather much, and if they're heart cedar, you'll smell them for years. From just the right angles, you'll be able to see daylight through the spaces between shakes, but rain will have to travel uphill to get through. And if the roof is metal, it will sound wonderful in the rain.

Remember that the wind gets under porches. All that square footage and toil can easily sail off down the hill in the wind unless you fasten it well. Lag screws help. So do little pieces of angle iron (hand forged look good) to anchor the posts. Because you don't have the tremendous weight to hold the porch down as you do with the logs of the house, consider building a heavy bolt (half an inch or more) into your stone foundation piers. Bolt the sill down, and fasten the joists and roof uprights to it securely.

You may want to use rough-sawn floor joists here, or hewn beams. Or just round poles. But be sure to use rot-resistant wood, such as cedar or pressure-treated wood, and flatten the top surfaces of the poles for uniformity. An adze is fine for this, after they're in place. Watch your feet, or they may not stay in place. Pressure-treated joists will last much longer but, of course, cost more.

If you use sawn stock, a couple of 20-penny spikes through the band (2×6 or 2×8) into the ends of each joist will hold, or toenail into the sill. A ledger 2×4 along the front band will help, and another along the house sill log will support that end. Or mortise joists into the house sill; use a heavy porch sill and mortise into that too.

If you plan to do a lot of jigging on the porch, you may want it stronger. Use a heavy porch sill and log or heavy beam joists, and mortise or notch them in. Because porches were often added later, and rebuilt or replaced several times over the years, almost anything can be found on older houses. Properly built and maintained, however, your porch should last as long as the house.

Heavy decking — 1¼ inches, 1½ inches, or 2 inches thick — is best on the porch. We sawmill ours out of white oak if available, and nail it so the shrinkage leaves cracks. Rain doesn't stay between the boards and rot them that way, and you can also keep an eye on all the exciting events that go on under the porch: rooster fights, skunk standoffs, infant mudpie sessions, snakes contemplating cabin entry. Even with cracks, it is necessary to slope the porch floor (about an inch in six to eight feet) to get the rainwater off before you slip on it, liquid or frozen.

To minimize cupping, we sometimes lay the boards out to season. And of course they cup. Then we turn them over and let the cupping straighten out before we nail them.

If you deck with tongue-in-groove flooring, it will, despite paint on the top and bottom sides, rot eventually. We did one in rare chestnut that won't rot for a long time. We do some in that ugly, green, pressure-treated wood. If you're decking solidly, do slope the floor away from the house.

Porch steps will probably be out in the weather, even with generous eaves. If it's a high porch, use two-inch sawn stock of oak, cedar, or cypress for the steps. Walnut or maybe Osage orange would be perfect, and your grandchildren could grow old on those steps. I use big blocks of stone a lot, and, really, a low porch needs no more than a couple of nice flat rocks as steps. Don't spoil the whole thing with concrete steps, and don't get too fancy. Marble and wrought iron are a little too much.

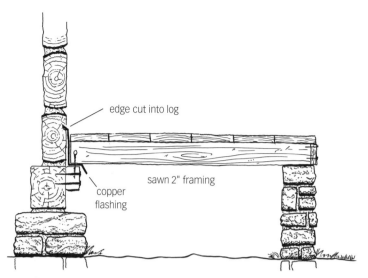

edge cut into log

sawn 2" framing

copper flashing

Porch floor joists may be attached to the house with a spiked ledger strip. Use treated wood here or a rot-resistant wood. Always use flashing when building up against a log wall.

This deck at Wintergreen is entirely of pressure-treated wood. Copper guttering minimizes water splashed onto the logs. The deck is stepped down at the far end to a hot tub.

Often, additions to a log house have a specific purpose. Most are post and beam or frame. Below, the rear addition serves as the kitchen/dining area for one house. The stone and stucco end of our house (left) has a log addition that serves as my wife's quilting room (also used as a guest room), a large bathroom, a storage attic, and the laundry room.

The western addition to our house was built by adding this 40-foot timber-frame house. The width of this house matched the log portion of the original log house. The 18-foot joist span is braced with a central oak sleeper, allowing smaller and less expensive floor joists. The exterior of this section was covered in stucco with stone at the end to balance with the stone at the eastern end of the house.

Building codes require railings if the steps or the porch are as much as 30 inches high, but you may want them anyway. Porch rails are a delight. I know of many in the mountains with carved initials and dates in them, and corners worn round from leaning, sitting, and propping up feet on them. My favorite rail is a rounded-corner 3×4 of heart cedar, split out and finished with plane or drawknife. Rails were usually mortised into the posts at about a 30-inch height. (Code, though, requires 36 inches.) These mortises were cut with a two-inch auger and a chisel.

Two holes were bored, one just over the other, and the wood left in the corners of the figure 8 was taken out with the chisel. Do this before you spike down the posts so you can drive the rails into the mortises at both ends. Code requires a maximum of four inches between rails or spindles, so plan a lot of detailed work here, important for a raised porch.

Really old railings featured spindles about one inch square turned one eighth with corners out, diamond shaped. This means mortising into either the top or the bottom rail to hold the spindles steady, but the other end can be drilled and toenailed with galvanized finish nails.

If yours is a high porch, bring the rail to the top of the steps, then join it to another for the step rail. The top post of the step rail will probably need no bracing, with rails into it from two sides, but the bottom post, or newel, will. If you don't mind a little shakiness, just spike it to the step framing stringer. I like to extend the bottom tread past the newel post and angle-cut or mortise a brace. Sometimes I forge an iron brace. For stone steps, I set a metal or treated post into concrete in the ground.

A gutter will keep water from dumping on you and from splattering on the steps as much. Porch guttering materials must match those of the rest of the house. Guttering and drainpipe are especially important to keep rainwater from ruining the porch support and flooring.

Remember to build your porch for use. If yours has a good view, spend a lot of time here appreciating it. Screen it against insects and other unwelcome creatures. Hold mountain music sessions here. Wear the porch into mellowness; properly used, it will have some effect on you too.

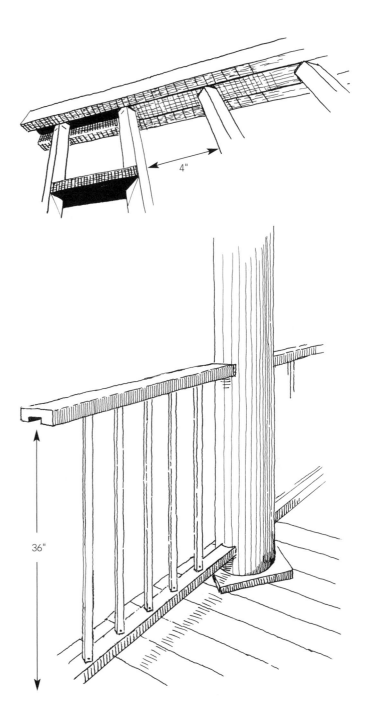

Porches higher than 30 inches off the ground require rails and spindles, strongly braced. Building codes require railings 36 inches high and spindles a maximum of 4 inches apart.

Lean-tos and Other Additions

These additions were just what their name implies. They leaned against the cabin and were almost always added later; they were usually made of sawn boards. When that long-suffering pioneer wife finally got her egg money together to buy a cast-iron cookstove, it often went into the lean-to. It kept the heat out of the main house in summer, and it was actually a modification of the old kitchen out back in a separate building. In a lean-to, this didn't do much for fire prevention, but it made more room.

The lean-to is still a good place for the kitchen and bathroom. It lets you believe that these modernizations could have been added later, even if your whole log house is new. There need be nothing very modern looking in the main house if these essential rooms are in the lean-to. Also plumbing is tidier in a frame "water wall," and electric wires hide well in stud construction.

Log cabin kit suppliers have come up with all sorts of ells, wings, and spacious goodies to uncramp the basic cabin, but nothing looks as homey as a board-and-batten or clapboard lean-to nudging the rear wall. I like them across the back, inset a foot or more to avoid covering those dovetail notches.

Another good-looking addition is a stepped-down room off the gable end, with the same roof pitch. Ours is a one-and-a-half-story stone addition built onto the two-story log section.

For the conventional, framed lean-to or end addition, spike 2×4 or 2×6 studs up the log wall after the logs shrink and settle; set sills on stone foundation corners, with flashing between; mortise or spike in joists; and you're on your way. Better add another stone pier under the back door for extra support at this heavy traffic point, or build a continuous foundation. Subfloor the joists as in any construction, insulate under, and put up your stud-wall framing. Tie rafters into house rafter ends as you did with the porch. Deck, insulate, then roof just as with the main house, and finish with guttering and drainpipes to match the rest of the house.

Do set in some angle bracing to keep this room from leaning. Modern builders use a sheet of plywood at the corners, covering the rest with fiberboard. You'll be happier, and so will I, if you angle-cut 2×4s and incorporate them into the stud walls. To insulate, cut the pieces to fit around these angles. And if you insulate under the floor with piers, use screen or rat wire to keep creatures from stealing it for nests.

Without this wall angle bracing, years of wind and wear will tilt things crazily. I know lots of old houses with their walls playing dominoes. Of course, with angled sheathing under the siding in the better early frame buildings, the bracing wasn't necessary. But with insulation, a double outer wall is really a waste of materials. Another inch of wood won't do nearly as much toward holding heat (in or out) as an inch of foam or fiber.

In building a hewn-log house, the procedures for installing windows and doors, inner wall covering, wiring, and plumbing are exactly the same as they are for conventional modern construction. I suggest board-and-batten outer walls as a visual break from the horizontal lines of the logs, just as for gable end covering. I use random-width boards, from 8-inch to 12-inch, with 3-inch battens. However, in the East, tradition was (and still is) to use clapboarding, butted to thick window and door trim and to corner boards.

We often build additions of timber-frame construction. That means we use fewer, heavier posts and horizontal beams for the major framing, filling in with a wall between to let the beamwork show — inside and, sometimes, outside too. It's a compatible mix, and because timber framing was common until about the Civil War period, it is often an authentic combination of building styles.

The one-and-a-half story timber-frame kitchen was constructed from a complete post-and-beam house recycled as a wing to our house. The stone-faced addition, with a loft office, was set back one foot from the corners of our log section. The dormers and bay window, with window seat, were framed later. The two-story raised porch is shown under construction and completed.

Porches and Additions Checklist

- ☐ Design a porch roof slope compatible with your house design.

- ☐ Frame the porch roof at the same time as the main roof.

- ☐ Secure the porch so it won't blow off in a high wind.

- ☐ Slope the porch floor (if decked solid) away from the house 1 inch in 6 feet to shed water.

- ☐ Set in angle bracing to keep a lean-to or an addition from leaning.

- ☐ Install gutters and downspouts on every porch and addition.

CHAPTER FIFTEEN

Lofts, Utilities, Finishing Interiors

THIS BOOK IS NOT ABOUT DECORATING, or household amenities, or standard carpentry. It is about dealing with a very specific kind of human abode and about creating space for living. Nevertheless, it is the ability to finish a house well that will ultimately make for a livable hewn-log house in the 21st century and beyond.

About space — you'll end up using every foot of it you can build into your cabin. Somehow our needs keep expanding, and every cranny can (and should) be utilized. Of course, certain things need to be considered early in the planning for the construction or restoration of a log house. For example, you need to use the longest logs possible in bedrooms and other rooms that are lived in the most, so that you can have adequate bed-wall space or space for other furniture. You will want to address traffic patterns for your home. You will want to leave adequate space for closets in the size of your choice, or for that antique chiffonier you bought for future use in your log house.

In addition to decisions about how space should be used, you'll also need to plan carefully for bathroom and kitchen utilities so that they will be accessible and frostproof. Extra care also needs to be taken in locating electrical wires and pipes for plumbing; hewn-log construction requires that the location of these be well planned. These decisions — and hundreds of other practical and aesthetic ones — need to be made for the construction of any log house.

Lofts

These delightful spaces exist any time you have a sloped roof above a ceiling. But too often there's only a cramped space under the ridge and lots of wasted cubic footage where the roof slopes to the walls.

In a log house, a usable loft is simply a matter of two or three more logs per wall and a ceiling strong enough for the upstairs floor. Well, it isn't *that* simple, as we pointed out in the chapter on roofs, but almost.

Of course, the higher the log walls (knee walls) above the ceiling, the more space upstairs; but there's a logical limit. Too many logs give you a tall, skinny, and funny-looking house. You'll recall that the top course of logs should not be cut into or otherwise weakened for windows. So logs should stop at the upstairs windowsills, or go on above the windows at a reasonable level, for a full two-story house.

Three feet or so of logs, plus a 45-degree roof, gives you 6 feet of height 3 feet out from the wall. Got that? You can shove a bed into that 3-foot height, or build

Loft space available with a three-foot knee wall and a 12/12 roof pitch requires creative planning. Doors sometimes must be clipped. Knee-level windows provide light, as do those in the gable above the top log.

closets that bring walls out to head height. If you put in an upstairs bathroom, shift things so that essential plumbing is located where there's headroom.

Any less than three logs up cuts your usable space drastically. One log up means you have to stay 5 feet out from the wall to clear 6 feet. In a 16-foot house, that's only 6 feet of standing room down the middle, and you may not have that as clear space if you want king-post support in the middle of the floor. Again, codes specify the knee wall or overall height required in livable lofts.

Lofts are almost necessarily dark, because you have window space only at the ends. There are no skylights. I build dormer windows in those larger houses that can handle them in good proportion. So did the better early builders. But dormers in a small cabin must be done tastefully. Then, they're delightful.

Dormers are fine in a 50-foot dogtrot, and they give a feeling of space. They're tricky to build, and mean lots of flashing and lots of weird angles to figure. They need rafters, decking, flashing, roofing, walls, the windows, and interior finishing. I love them.

Once, faced with the problem of locating an upstairs bathroom (it had to be against a wall so the pipes wouldn't come down into the middle of the living room), I put the shower in a dormer window. The owners loved it; the house was in the deep woods, and there were relatively few prying eyes around.

We've talked about bracing the ceiling joists with king or queen posts from the rafters. If the house is 20 feet or more deep, it's a good idea to use king posts and build a dividing partition around them. That gives you two rooms upstairs, each with its sloping roof. Or, for smaller houses or houses with no upstairs log walls (limiting space), use the queen posts to define a space with headroom down the center.

Because a 45-degree roof gives you so much height at the peak (13 feet in a 20-foot-deep house with 3-foot knee walls), it's often a good idea to put a ceiling at collar tie height and have all that peak for storage. This also cuts down on the area you must insulate and allows a flat ceiling here to use conventional insulation above.

Of course, you may want to leave out the ceilings altogether. You may want no upstairs; no ceiling joists; just all that space, soaring off up there. Unfortunately,

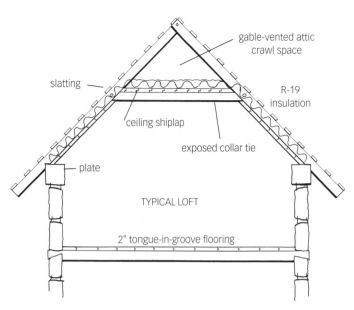

Most early cabins had knee walls, making almost the entire upper level usable. That is still a good idea in modern houses.

This loft made a bright, cheerful bedroom. The space shared a walk-in closet and small but well-appointed master bath.

that's what the heat will do also. Aside from that, and the 50 percent reduction in floor space, this is fine. I don't recommend it, and evidently neither did the pioneers, who always seemed to have more warm bodies than space to stow them.

It would seem that if a 45-degree roof gives you all this space, a steeper one gives you more. It does. But it's a lot harder to build a steeper roof, and takes longer rafters and more of every material. And it looks weird. Whereas lower pitches (8/12, 6/12) were frequent, especially in taller houses, those steeper than 12/12 were almost nonexistent.

Kitchens

Often we recommend that the kitchen and bathrooms be placed in the frame or post-and-beam additions of log houses, since it is easier to hide plumbing in frame walls. However, a kitchen can also be built in a log room. The kitchen is such a personal space that it's hard to generalize here. Our clients have wanted everything from open shelves to granite countertops (sometimes both). Any kitchen design should minimize space among the three most functional centers: stove, sink, and refrigerator. Log walls affect kitchen layout little, unless pipes from upstairs must come down here. Sometimes this is unavoidable, so the pipes are hidden in chinking space, in vertical chases along the wall, in horizontal chases between beams, or in cabinets.

A colorful addition to a kitchen is a cooking fireplace, with crane, Dutch ovens, and kettles. This can sometimes be built back-to-back with the main fireplace, in the same chimney. The Sam Black Tavern restoration was done this way, a duplicate of the 1769 original. We often wondered how the cooks worked, bent over those fireplaces. Today, you can keep the ambience but make the kitchen fireplace more pleasant by raising the hearth to the desired height.

Installing Utilities

Electricity is the utility you'll need (or miss) most. I favor cutting our dependency on this form of energy as much as possible, since we have been grossly spoiled in the past two generations by gadgetry that plugs in.

We have friends who use bottled gas for their refrigerator, water heater, and small cookstove (in summer); they use a wood cookstove in winter, and light with kerosene. They have a battery-powered tape player. But they do keep a freezer and a washing machine at a friend's house.

I grew up in a log cabin where we had no electricity until I was 15. We carried water, sometimes chilled things in a spring, lighted the house with kerosene lamps, listened to a battery-powered radio, and cranked a Victrola. As I remember it, our first purchases for electric usage were lights, a refrigerator, and a water pump.

That's one plateau. You still must cook, but I learned my limited culinary skills at a woodstove, and several of my friends now use wood from October to May. Hot water can be had in reasonable quantities from the reservoir in the cookstove, and you *can* dip in your creek in summer.

If you choose to electrify your new or restored log house, there are a few things to know. Even if you're outside the cities that have strict codes, your local power company will discourage the kind of offhand wiring that could burn down your house. They pay specialists to provide information and suggestions, so go see one.

Find out if the company will even extend its line to your site. Then, if you want to bury the wire, get its specifications. Depending on the usage you and the company's specialist anticipate, you'll need perhaps a 100-amp or even a 200-amp entry panel, which connects to the meter base. Different-size conduits are required for different amperage.

It looks best if the electric box is mortised into the log face, because the fixture or outlet fits flush and vertical. This work takes more time and requires a lot of drilling and chiseling, but the end product is very satisfying. Here the box is chiseled into the exterior of the log and will hold a lighting fixture next to the back door. In this photo, the wiring is encased in conduit. We do not normally use any wire other than Romex, since conduit is necessary neither to meet building codes nor for other criteria. However, the electrician preferred this method and the owner approved. Running Romex wire between and through drilled logs works just as well, as long as all the building codes are met.

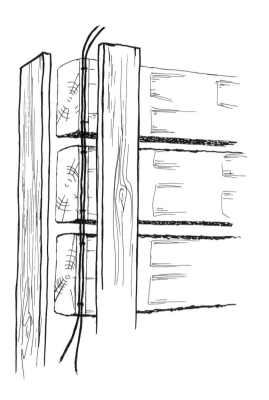

Electric wires can be boxed in under window and door trim (as shown here) but are better installed in holes drilled through the logs during assembly. Get specifications from the county inspections office.

Your panel will have a main disconnect switch, which lets you shut off everything and avoid getting burned. Every electrician I know has at least once turned off the wrong breaker switch and discovered it by burning off the end of his screwdriver on a "dead" circuit.

Three-wire or two-wire-with-ground, 12-gauge plastic-covered cable (Romex) is standard for house wiring. Staple it with the wide, padded staples made for that purpose. Run circuits wherever possible between logs.

Vertical wiring goes up best through *drilled logs,* but can be done nicely in a space left by beveling the log ends, to be covered by facings at windows and doors. Metal plates must be put over these, so nails cannot be accidentally driven into the wires later. Sometimes we bore two intersecting holes through a log to escape a cul-de-sac at a corner. And with a raised wood floor, lots of the wire can be strung underneath to come up through maybe one log to the outlets. The outlets look best mortised into the log faces, which means drilling from the chinking crack into the recessed mortises. It's hard to get outlet boxes set flush

in the chinking. Most locations let you, the owner, wire your own house, but require that an electrician have a master's card.

We always drill vertical one-inch holes for the wire at the doorways, down into the crawl space and up past switch height, on up through the top plate for access into the ceiling or roof. For overhead lights, we often route out a groove in a beam before we floor upstairs, and drill through for the fixture.

Power companies frown on the practice of embedding wiring in concrete, on the theory that acids will eat it, or at least the insulation off it. Maybe they're right; I have no idea what chemicals badger each other in those combinations. Metal conduit can be used here, however.

It's obviously easier to properly wire a log house (or any other house) before it's complete, so I advise you to go ahead with it. Even the hardiest of my forest-dwelling cohorts entertains schemes to produce his own power someday, and even minimal electricity flows better through adequate circuits.

Do get specific instructions before you wire. It's expensive to have to do it over. There are numerous how-to books on the subject available from the library or bookstore. And you can take courses in residential wiring at the local vocational/technical school.

Of course, this could be the ideal place to spend your dollars on a professional. He'll cost you $40–$60 an hour, however.

Wiring for Telephone

You may miss the telephone the least or the most, depending on your hang-ups and whether you operate a business, as we do. Linda could not run her publications operation or my construction business without a phone, and wouldn't want to run her life without one. And now faxes and modems are routine in even the smallest offices.

Very simply, the telephone company has to be able to reach you if you want the joys of conventional long-distance conversations. The company will just about always come in on an electric company pole, and will let you bury its line in the same ditch as the electric cable, at a reasonable 12-inch distance to avoid static. Or it will bury the line for you at no cost.

Here is what NOT to do when retrofitting wiring in a cabin. The owner tried to shortcut the process by installing wiring without putting the wiring into the chinking crack. That would have required rechinking the whole house — which it definitely needed anyway. This is neither safe nor attractive. In the next example, the wire has been installed properly, but the owner has not built the chase yet.

The pioneers solved the problem of communication in a most pleasant way: They went visiting often. Country churches, square dances, quilting bees, house-raisings — these gatherings were the news centers, of course, and so were the individual visits to borrow fire, lard, or gunpowder.

A relatively recent rural community news bearer was the late Ted Richmond. He carried books in a pack from his Wilderness Library to hill folks in Newton County, Arkansas. Until the early 1950s, Ted would set out regularly from his round-log cabin on the north slope of Mount Sherman, bound for other remote holdings with books and magazines. Ted lived alone, devoting his life and energies to visiting the hill people, some of whom never quite accepted him. We shall probably not see his like again: the visitor on foot, arriving to cut wood for the ill, to visit long hours, to bring news and the magic of print.

His cabin is falling into ruin today, and rats have shredded the books and magazines that covered the floor and jammed the loft. He was considered something of an oddity, choosing to live as he did when almost everyone who could was leaving the hills around him. It will not be many years before the mountains swallow up all traces of him and his work, as well as those of so many of the folks he worked with. In the years since I visited this site, I know the decay process has gone a long way.

Boxed-In Areas

You may have to enclose pipes from an upstairs bathroom in a chase, which is simply a boxed-in shaft. Or you can build in a closet to hide the pipes. Because plumbing fixtures such as toilets and showers need drain traps, you'll also have to box in between ceiling beams to cover these. Try not to cut through beams for big pipes; it weakens them. We usually set this boxing up between the beams as far as possible to let part of the beam show, then insulate around the pipes to cut the noise. A cast-iron main drain here reduces noise of upstairs flushing

Upstairs you may have irregular shapes to finish if you have inner roof surfaces, low roof angles, dormers, or odd-shaped windows and doors. Here again we use shiplap paneling, unless an owner wants drywall. In that case, we subcontract this part of the job. (The people who work for me do so to get away from that kind of building.)

Lighting

Lighting is a major consideration inside your cabin, and is often difficult to do well. A constant complaint is that log cabins are dark. During daytime, you rely on the number and placement of windows. However, nighttime is another matter. You do not have the advantage of white walls to reflect your light source. The dark wood will absorb much of any illumination. You will want to play every trick in the book to lighten your log room, including decorating with light-colored furniture.

Increasingly, builders and restorers are cutting out log space for more and larger windows. This weakens the structures and loses much of the traditional cabin appearance. However, with judicious planning, you can use larger or more numerous windows in selected walls. It is especially pleasant for the kitchen to be bright and cheerful during the day without artificial lighting. Keep in mind, though, that additions like porches will often darken even a room with good window placement.

Lighting fixtures are both decorative and functional items to accomplish specific tasks. You are not limited to using a fake wagon wheel hanging light. You have many options from custom to commercial fixtures. If you have strong preferences and some knowledge, you can make these choices yourself with careful shopping. Electric supply companies often have staff lighting designers available who can be of great help.

Most cabins have more beamwork than a normal house and this must be taken into consideration.

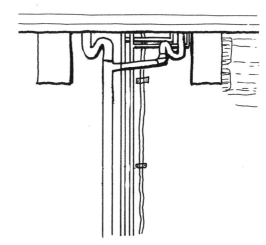

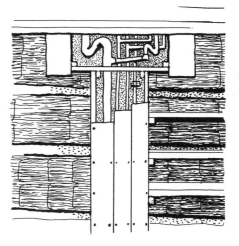

Boxing in plumbing pipes with an open-beam ceiling can be done with a chase. Recessing the boarded-in ceiling area leaves the beamwork exposed. Sound can be deadened with insulation.

Here, a chase is built against a log wall to house both plumbing pipes and electrical wires. Chases can be camouflaged by making them into cupboards or parts of bookshelves (below). HVAC chases can also be installed in a log house, but will require more careful planning.

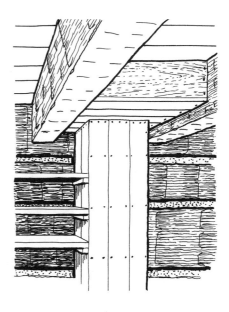

Track lighting, concealed behind an overhead beam, can wash over a stone fireplace nicely or become task lighting for a kitchen. Recessed lighting has its limitations in a log house. Since the ceiling is usually the underside of the upstairs floor, wall sconces and chandeliers work better.

As in any house, you can operate any lighting with three-way switches. When building a log house, you can install the wiring for as many three-way switches as you want. However, if you are installing a new lighting fixture in an existing log house, it is very difficult to retrofit any light switch. You can't just open the drywall and fish for wiring. So plan ahead.

For kitchens, where good light is mandatory, count on more light than usual. Task lighting as well as decorative lighting are personal preference. However, the kitchen is a workroom and requires better planning and more attention. Nothing is more frustrating than a dark, poorly lit kitchen. And nothing is more of a delight than a welcoming, functionally lit kitchen. Track lighting, over-the-counter lighting, and hanging fixtures all work well.

Lamps are very much a matter of personal taste in any house. You will need more of them to counter the nonreflective wood. Your individual needs will determine whether to use floor lamps and table lamps and where to use them.

Historically, cabin dwellers whitewashed the walls and ceilings to lighten things. Few of us today would

Lighting can be as simple or as elaborate as you choose. Decorators have done very creative things with some of our buildings and restorations. This chandelier is of forged iron and turned wood.

want to cover the logs this way. But painting the ceiling off-white between natural beams is an easy, subtle, and cheap way to lighten the interior and give the illusion of height. The white ceiling will not detract from the natural color of the beamwork or fight with the furnishings. The white ceiling will help reflect the room lighting, instead of absorbing it.

HVAC and Plumbing

Heating and air-conditioning are increasingly being used in log cabins. Because ducts can't be hidden in log walls, chases often must be used. Sometimes vents can open under stairwells or in corners behind fireplaces to be less obvious, but they must be functional: supply and return air passages. In a log cabin, we have had to place return units in odd places, above ceiling rafters and bathroom ceilings. So it can be done. Heating and air-conditioning contractors will design systems, and give you options for placement, size of units, and efficiency.

About the only applicable limits to plumbing in a log cabin apply to log walls: You can't hide pipes in them. Here, both vertical and horizontal chases are necessary. Again, use cast iron for the main drain from upstairs to deaden the sound of flushing.

Interior Finishing

You'll need interior walls for things like bathrooms, and to divide large areas. Traditionally, log builders used a "board wall" or a "box wall" of shiplap or tongue-in-groove boards anchored to floor and ceiling with no stud-wall framing. You could hear whispers through those thin, uninsulated walls, and if one room was heated and the next not, the heat flowed right through.

We come as close to conventional modern building here as anywhere in our work. We lay a sill 2×4 or 2×6, studs on it at 16 inches o.c., and a plate 2×4 or two over the top. Sometimes this can be nailed together on the floor and stood up, but sometimes it must be done in place, toenailing the studs in. Drill holes and install electrical wire and outlet boxes, then insulate, at least for sound deadening.

Wall covering is up to you. Drywall is cheapest, of course, but has a more jarring effect than plaster does. It is, however, virtually necessary if you want to use wallpaper. We use wide-board paneling, 1×10 or so, shiplapped and nailed up vertically. Then we paint it if it's new wood, or oil it if it's old or if we don't need a light color to fight the drabs. Labor is less for the wood, because there's no taping, spackling, or sanding. But the wood costs more than the drywall.

We like the wood wall because, even under the paint, you see grain, knots, and all the irregularities that are absent in that maddeningly bland drywall. Also, this is the look of the traditional interior wall. You can even hand-plane the wood for texture.

Ceilings

If you want to see heavy beam rafters, you can insulate between them and recess ceilings between to show part of the beams. Or do a built-up roof, as we discussed it. Usually we rafter with 2×10s and panel right against them with the wide boards. This doesn't show any rafter. Often we show heavy collar ties and take the ceiling right up to the peak, which can look good, especially if the collar ties are hand hewn. Sometimes, we apply original 3×4 rafters, pegged at the peak, under the ceiling boards. Purely ornamental, but nice.

This bedroom incorporates drywall with the log walls. Logs are irregular and difficult to interface with dimensional lumber or other building materials. Therefore, drywall, paneling, and trim boards must be adapted to fit the natural peculiarities of the log surface. The frame room partition wall in this bedroom required the drywall to be carefully scribed to the perpendicular log walls. In addition to normal use of drywall compound on the surface of the drywall, we often use paintable caulk where the drywall and logs touch.

Venting

Venting the roof is hard if you take the ceiling to the peak. It's necessary for air to move outside the insulation; otherwise, moisture builds up. With the ceiling across at collar-tie height, you insulate here and use gable vents to let the air move. With a peaked ceiling, you must use some sort of ridge vent. This is easy with asphalt shingles, but shakes or standing-seam metal don't work well with a ridge vent.

We've solved this problem in some houses by leaving a space above a lowered ridgepole and mounting a small cupola for venting. But this doesn't look right on a log house. I prefer to expose the collar ties, but I put a flat ceiling above them and use the gable vents. Another solution I've used is to space the roof decking as the early builders did, to allow air space next to the roofing. This lets the roof vent horizontally. When we do this, we screen the spaces at the eaves to keep out birds and mice. Also, if we've used soffits at the overhangs, we put in vents there so air can move up, then out horizontally.

Closets are often an afterthought in log house construction, but are vital for modern living. Pioneers used trunks or cupboards. Today, we often want walk-in storage. Here, a spacious cedar-lined closet base was built into the knee wall of a one-and-a-half-story loft bedroom.

The high dormer window provides pleasant light for the open kitchen.

Log Wall Finishing

Log walls need no finishing except to wirebrush the dirt off or change color from weathered gray to the mellow brown tone of aged wood. A coat of linseed or tung oil will bring out the color and grain. If you use grayed old logs, don't oil them or they'll go black. Don't use anything on the outside.

We often pressure-wash the old logs to get off whitewash, dirt, and graffiti. This leaves some fuzz, which can be taken off with steel wool. Age gives the rich color to wood that stains can't match.

Sometimes we're asked to cover log walls with drywall or paneling, which I refuse to do. If you don't want to look at log walls, you don't really want a log house.

These additions to your basic cabin make it the livable house you'll enjoy. They also constitute a lot of the overall construction time, and expense. Often, we do these one at a time, as we (or the owners) can afford them. After all, that's the way the little house–big house grew.

LOFTS AND FINISHING MATERIALS CHECKLIST

☐ Go three logs up above the ceiling to create usable loft space.

☐ Consider dormers to add light to loft space.

☐ Conceal electric wiring between logs before chinking.

☐ Hide plumbing pipes in boxed-in chases.

☐ Design and install more lighting to overcome dark walls.

☐ Try wide-board paneling as a more aesthetically pleasing alternative to drywall.

EPILOGUE

YOUR LOG HOUSE EXPERIENCE will be a delicious and varied affair, no matter how involved you become. You may be fortunate enough to locate and acquire a silvering cabin hidden in some beechwood glade and touch it into life again.

Or you may begin with the raw earth, in the woods, at the edge of an overgrown field, or even just out of town — where there is room for you and your house to breathe. You will soon learn that a hewn-log house is more work and/or expense to build than just about any other structure. It will exact from you blisters for a lifetime, effort and skill you may not have known you possessed. Building it admits you to a rare and privileged brotherhood of craftspeople, steeped in a love of labor and very near to the living earth.

Or you may simply find that your sensitivities are tuned to the very fact that some pioneer houses still survive, and you may become aware of their existence. You may drive out of your way to find those that are perhaps deserted, falling into decay.

Spend some time there now that you better understand the wilderness call our fathers heeded, and the singing of pioneer blood in their veins. Listen for the sound of axes ringing in the frost of morning, and the echo of old lullabies, sung before worn hearths of timeless stone.

BIBLIOGRAPHY

Glassie, Henry. "The Appalachian Log Cabin." *Mountain Life and Work*, 39 (1963): 5–14.

—— *Pattern in the Material Culture of the Eastern United States.* Philadelphia: University of Pennsylvania Press, 1968.

Hutslar, D. A. "Log Architecture of Ohio." *Ohio History*, 80 (1971): 172–271.

Jordan, Terry G. *American Log Building: An Old World Heritage.* Chapel Hill, N.C.: University of North Carolina Press, 1985.

Kniffen, Fred. "Folk Housing: Key to Diffusion." *Annals of the Association of American Geographers*, 55 (1965): 549–577.

—— "On Corner Timbering." *Pioneer America*, 1 (1969): 1–8.

McCanse, Ralph Alan. *Titans and Kewpies: The Life and Art of Rose O'Neill.* New York: Vantage Press, 1968.

McDonough, Nancy. *Garden Sass: A Catalog of Arkansas Folkways.* New York: Coward Geoghegan, 1975.

Montell, William Lynwood, and Michael Lynn Morse. *Kentucky Folk Architecture.* Lexington: University of Kentucky Press, 1976.

Roberts, Warren E. "Some Comments on Log Construction in Scandinavia and the United States." In *Folklore Today: A Festschrift for Richard M. Dorson*, edited by Felix J. Oinas, Linda Degh, and Henry Glassie. Bloomington: Indiana University Press, 1976.

—— "The Whitaker-Waggoner Log House from Morgan County, Indiana." In *American Folklife*, edited by Eon Youder, 185–207. Austin: University of Texas Press, 1976.

Sloane, Eric. *A Museum of Early American Tools.* New York: Random House, 1964.

Smith, J. Frazer. *White Pillars.* New York: Bramhall House, 1941.

Stahle, David, and Daniel Wolfman. "The Potential for Tree-Ring Research in Arkansas." *Field Notes: The Monthly Newsletter of the Arkansas Archaeological Society*, no. 146 (February 1977): 5–9.

Vogt, Evon Z., and Ray Hyman. *Water Witching U.S.A.* Chicago: University of Chicago Press, 1959.

Weslager, C. A. *The Log Cabin in America.* New Brunswick, NJ: Rutgers University Press, 1969.

Wigginton, Eliot, ed., *The Foxfire Book.* Garden City, NY: Doubleday, 1972.

Wilson, Eugene M. *Alabama Folk Houses.* Montgomery: Alabama History Commission, 1975.

—— "Some Similarities Between American and European Folk Houses." *Pioneer America* 3 (1971): 8–14.

ADDITIONAL READING

Bealer, Alex W. *The Log Cabin.* Barre, MA: Barre Publishing, 1978.

Gott, Peter. *Log Buildings and Workshops.* (Pamphlet.) Marshall, NC: Self-published, 1990.

Mackie, B. Allan. *Notches of All Kinds.* Prince George, British Columbia: Canadian Log House, 1977.

McRaven, Charles. *The Blacksmith's Craft.* North Adams, MA: Storey Publishing, 2005.

—— *Stonework: Techniques and Projects.* North Adams, MA: Storey Publishing, 1997.

—— *Building with Stone.* North Adams, MA: Storey Publishing, 1990.

Mitchell, James. *Short Log and Timber Building Book.* Point Roberts, WA: Hartley & Marks, 1984.

Langsner, Drew. *A Logbuilder's Handbook.* Emmaus, PA: Rodale Press, 1982.

Peterson, David, and Peter Gott. *Building the Traditional Hewn-Log Home.* Hendersonville, NC: Mother Earth News Incorporated, 1987.

Phelps, Hermann. *The Craft of Log Building.* Ottawa, Ontario: Lee Valley Tools Ltd., 1982.

Tempel, John I. *Building with Wood.* Toronto, Ontario: University of Toronto Press, 1967.

Thiede, Arthur, and Cindy Teipner. *Inside Log Homes: The Art and Spirit of Home Decor.* Salt Lake City, UT: Gibbs Smith Publisher, 2003.

—— *The Log Home Plan Book.* Salt Lake City, UT: Gibbs Smith Publisher, 1999.

—— *Hands-on Log Homes: Cabins Built on Dreams.* Salt Lake City, UT: Gibbs Smith Publisher, 1998.

—— *The Log Home Book.* Salt Lake City, UT: Gibbs Smith Publisher, 1995.

—— *American Log Homes.* Salt Lake City, UT: Gibbs Smith Publisher, 1986.

Glossary

adze — smoothing tool with the blade perpendicular to the 3-foot handle

angle brace — any timber bracing a corner, across that corner

auger — hand boring tool

barn sash — single sash, simply a wooden frame with glass panes

batter board — horizontal boards nailed to stakes, to which foundation layout strings are attached

beaded beam — usually a ceiling joist ornamented with a groove near each rounded corner; beading was done with a shaping plane

bird's mouth — a notch in each rafter at the corner of the plate

bird-stop board — board fitted between rafters where they pass over the plate, extending the wall to the roof

block and tackle — series of pulleys roped together to give a mechanical advantage for lifting or pulling

board and batten — vertical siding of wide boards, cracks covered by narrow strips

board foot — the standard way to price and measure lumber using 12×12×1 inches as the base calculation. For example, at $1 a board foot, a 10-foot-long board 1 inch thick and 12 inches wide would cost $10 and a 6-foot-long board 2 inches thick and 12 inches wide would cost $12

bucks — vertical boards nailed or pegged to log ends to create window and door openings; the measurement for the bucks must match the rough openings of the windows and doors to come

bob truck — truck with one frame, not a truck-and-trailer unit

broadaxe — wide-faced axe beveled on one side, for hewing

bulkheads — boards set across the foundation ditch to contain and level poured concrete

bunching — groupings logs for loading

butting poles — part of the roof, poles laid into notches in the top end logs to support weight poles and knees, which held the early shakes in place when nails were not available

cant hook — a pole with a hook near one end for rolling logs

catted chimney — one built of dried clay rolled into "cats" laid within courses of small pole pens

chalk line — string loaded with chalk for making lines

chamfer — beveled corner of a timber

chinking — the filling between logs; mortar, or short split boards and clay

clapboards — overlapping siding boards laid horizontally

collar tie — a horizontal beam between rafter pairs near the roof peak

concrete — aggregate of cement, sand, gravel, and water, usually steel-reinforced

conduit — pipe for electric wire

corner-notched — log ends fitted at the house corners

course of logs — corner-notched logs at one level around the house perimeter

crosscut saw — usually a coarse-toothed saw for use across the grain

cross-haul — a system of chains or ropes for rolling material up skids

dado blade — circle saw blade set to weave and make a wide cut

decking — roof covering onto which shingles are laid

dogtrot house — a style originally two pens separated by a breezeway, with a common roof and usually fireplaces at opposite ends

double-hung sash — a sliding window, with halves above and below

dovetail notch — an angled notch at the end of a log beam or timber, cut straight in and widening to the end

drawknife — a smoothing tool; two-handed shaving knife drawn toward the user

drystone foundation — stones stacked without mortar to form a base on which to build

eaves — overhang of the roof

effluent — liquid waste

ell — an addition usually at right angles to a house

entry panel — electric switch box with fuses or breaker switches

facing — also called trim; finishing board nailed to the jamb at the interior and exterior wall surface; used around windows and doors (not the same as bucks)

field lines — septic tank drain lines buried level to dissipate runoff liquid

flashing — sheet metal used at the roof peak, in roof angles and under the shingles to prevent leaking

floor joists — beams mortised or notched onto the sills, onto which flooring is laid

flue tile — clay pipe used to line chimneys for a cleaner, smoother surface

footing — wide masonry support in the ground on which foundation is laid

froe — L-shaped tool for splitting or riving shakes or staves from wood blocks

full 2×4 — timber of that measure; lumber company offerings usually measure 1½" × 3½"

gable — the triangle wall space enclosed by the roof at the ends of the house

gin poles — boom of two poles braced by a cable or another pole for lifting

girder — heavy bracing beam

GPM — gallons per minute, the measure of water flow

half-dovetail — a log notch with only one outward slope to the end

heat box — dual-wall metal fireplace liner that improves heat circulation

hewing — flattening or squaring timber, usually with the broadaxe

hewing dogs — iron stays driven into logs to hold them steady for hewing

house-raising — the neighborhood gathering to build a house cooperatively

hydraulic ram — pump that operates on the force of water moving downhill, which transports a part of that water up a higher hill

joist — timber on which flooring or ceiling is laid

juggles — heavy chips cut by a broadaxe

kerf — the path left by a saw cut

keystone — V-shaped stone at the center of an arch

kiln-dried lumber — that with the moisture baked out of it

king post — vertical support for the joist down from the rigid roof peak to the midpoint of the joist

knee — a roof element, short timber laid vertically between weight poles

knee wall — partial wall between upstairs (loft) floor and roof

lag screw — heavy screw for wood with bolt head

laths — spaced boards or slats that serve as a base for covering

lean-to — three-sided addition to a building, with shed roof, usually in the rear

ledger — see *nailer strip*

lintel stone — stone spanning an opening such as a fireplace

loft — space upstairs under the roof, attic

log sitter — cooperative assistant to keep log steady

masonry cement — mixture of lime and cement that is mixed with sand and water for mortar

metal lath — screening used for reinforcing plaster

mortising — joining two timbers by stepping down the end of one (the tenon) to fit into a hole cut into the other (the mortise)

muntin — frame divider between glass panes in a window

nail set — punch for sinking finishing nails below the surface of the wood

nailer strip — (also ledger) light timber fastened to wall to give support for joist ends, flooring ends

newel — the bottom post to which the stair railing and, sometimes, the stringer are attached

nogging — a filling of clay or brick in walls or floors

notching — cutting away pieces of wood to enable a beam or log to fit to another

o.c. — stands for "on center"; a repetitive measurement of the distance from center to center of framing pieces (maybe measured from matching edges, with location marked); i.e., roof rafters are laid out on 24-inch centers.

parge — to plaster with cement

pegging — pinning, as with trunnels

pens — squares or rectangles built of corner-notched logs

pier — column used as foundation

pinned — attached with wooden pins or trunnels

pit saw — vertical sawmill, powered by steam, water, or muscle

plate — the top timber or log of a wall to which the rafters attach, usually wider than the wall logs

plumb bob — pointed weight on a string for locating exact vertical point or horizontal distance on a slope

post and beam — type of construction involving widely spaced, heavy uprights set on heavy sills with heavy plates above

pry bar — metal leverage tool, crowbar, pinch bar

puncheons — heavy flooring timbers, originally halved logs with the flat sides up

purlins — lengthwise roof timbers to which shakes are nailed

push pole — temporary extension for raising timbers, trusses, rafters

queen posts — pair of vertical supports down from each end of a horizontal beam (collar tie) near the roof peak

rabbet — inset groove, as for setting glass in windows

rafters — beams set at an angle from the plates at the tops of the walls to the ridge, to which the roofing is attached

rafter truss — triangular structure of two rafters and a horizontal brace or chord; several make up the roof structure

ridgepole — lengthwise timber supporting the upper ends of rafters, rarely used in early houses

risers — the vertical surfaces of steps

riving — splitting, as in shakes

riving horse — two-branched support for riving or splitting shakes and staves

saddlebag house — a style, originally two pens together at the gable with a fireplace between

scab — temporary board brace

score-hack — the perpendicular cuts along a log to the depth it is to be hewn

septic tank — container in which sewage solids are broken down by bacteria

shake — a thin split board used as a shingle

shaping plane — smoothing tool with a shape or design in the cutting blade

shaving horse — a clamping device to hold wood for smoothing with a drawknife

sheave — enclosed pulley for rope or cable

shingle — thin roof-covering board

shiplap — shaped-edge boards that fit overlapped, also called half-lap

shutters — wooden or iron window covers

sills — heavy base logs that sit on the foundation and onto which the floor joists and walls are set

skidded — towed along the ground, as in logging

skids — inclined poles or timbers for loading or raising heavy material

slats — staves, often described as the widely spaced boards that shingles are nailed to

sleeper — heavy beam bracing joists; also called a girder or summer beam when the beam supports the ceiling joists

slick — a long-handled wood chisel used as a smoothing tool; should be pushed, never stuck

smithy — blacksmith shop, often confused with the smith, who works there

snow birds — metal devices clamped to the roof to keep the snow from sliding off the roof and damaging gutters, plantings, and people; the devices allow the snow to melt in place

spanner — stringer, any timber reaching from one support to another, spanning an opening

spindle shaper — mechanized, circular shaping tool

square — carpenter's tool for determining 90-degree and other angles

staves — split boards longer than shakes

stile — steps over a fence

stone hammer — tool for shaping stone

stringer — any timber used to span a distance, usually the side timbers in stairs

stud wall — wall framing of light vertical timbers or studs attached to a sill, capped by a plate

summer beam — sleeper, joist support for a ceiling

tenon — the stepped-down end of a timber that fits into a mortise

30-penny spike — heavy nail, a size designation, originally 100 could be bought for 30 pennies

tongue-and-groove lumber — also called center match, t.i.g. lumber, shaped at two edges to interlock

tread — the horizontal surface of a step

trunnels — wooden pegs, or treenails, used to fasten beams together

turkey-feather roof — shingle or shake roof with the top row extended above the ridge

two-foot center — distance from the center of one joist, stud, or rafter to another

water witcher — person skilled in finding water beneath the ground

weight poles — a roof element, poles holding down shakes before nails were available, often used with knees

whipsaw — a man-powered vertical sawmill

wings — additions to the sides of the house

INDEX

Page numbers in *italics* indicate illustrations or photographs.